CW01018811

Studies in Computational Intelligence

Volume 772

The series "Studies in Computational Intelligence" (SCI) publishes new developments and advances in the various areas of computational intelligence—quickly and with a high quality. The intent is to cover the theory, applications, and design methods of computational intelligence, as embedded in the fields of engineering, computer science, physics and life sciences, as well as the methodologies behind them. The series contains monographs, lecture notes and edited volumes in computational intelligence spanning the areas of neural networks, connectionist systems, genetic algorithms, evolutionary computation, artificial intelligence, cellular automata, self-organizing systems, soft computing, fuzzy systems, and hybrid intelligent systems. Of particular value to both the contributors and the readership are the short publication timeframe and the world-wide distribution, which enable both wide and rapid dissemination of research output.

More information about this series at http://www.springer.com/series/7092

Anis Koubaa · Hachemi Bennaceur
Imen Chaari · Sahar Trigui
Adel Ammar · Mohamed-Foued Sriti
Maram Alajlan · Omar Cheikhrouhou
Yasir Javed

Robot Path Planning and Cooperation

Foundations, Algorithms and Experimentations

Anis Koubaa
Prince Sultan University
Riyadh
Saudi Arabia

Hachemi Bennaceur
College of Computer and Information Sciences
Al Imam Mohammad Ibn Saud Islamic University
Riyadh
Saudi Arabia

Imen Chaari
University Campus of Manouba
Manouba
Tunisia

Sahar Trigui
University Campus of Manouba
Manouba
Tunisia

Adel Ammar
College of Computer and Information Sciences
Al Imam Mohammad Ibn Saud Islamic University
Riyadh
Saudi Arabia

Mohamed-Foued Sriti
College of Computer and Information Sciences
Al Imam Mohammad Ibn Saud Islamic University
Riyadh
Saudi Arabia

Maram Alajlan
College of Computer and Information Sciences
Al Imam Mohammad Ibn Saud Islamic University
Riyadh
Saudi Arabia

Omar Cheikhrouhou
College of Computers and Information Technology
Taif University
Taif
Saudi Arabia

Yasir Javed
College of Computer and Information Sciences
Prince Sultan University
Riyadh
Saudi Arabia

ISSN 1860-949X ISSN 1860-9503 (electronic)
Studies in Computational Intelligence
ISBN 978-3-319-77040-6 ISBN 978-3-319-77042-0 (eBook)
https://doi.org/10.1007/978-3-319-77042-0

Library of Congress Control Number: 2018934389

Printed on acid-free paper

This Springer imprint is published by the registered company Springer International Publishing AG part of Springer Nature
The registered company address is: Gewerbestrasse 11, 6330 Cham, Switzerland

Preface

The objective of the book is to provide the reader with a comprehensive coverage of two important research problems in mobile robots, namely global path planning and cooperative multi-robots applications with a focus on multi-robot task allocation (MRTA) problem. As such, this book is organized in two major parts: Global Path Planning, and Multi-Robot Task Allocation. The objective of the first part of the book is to respond to a research question that we have been investigating along the two-year period of the iroboapp project: considering the vast array of AI techniques used to solve the robot path planning problem ranging from evolutionary computation techniques (e.g. GA, ACO) to meta-heuristic methods (e.g. A*), which technique is the best? In this part, we first revisit the foundations and present a background of the global path planning problem, and the underlying intelligent techniques used to solve it. Then, we present our new intelligent algorithms to solve these problems, based on common artificial intelligence approaches, and we analyze their complexities. Different simulation models using C++, MATLAB and others have been devised. An extensive comparative performance evaluation study between the path planning algorithms is presented. In addition, we validate our results through real-world implementation of these algorithms on real robots using the Robot Operation System (ROS). The second part of the book deals with cooperative mobile robots. We focus on the multi-robot task allocation (MRTA) problem and we present a comprehensive overview on this problem. Then, we present a distributed market-based mechanism for solving the multiple depot, multiple travel salesman problem which is a typical problem for several robotics applications. A major contribution of this book is that it bridges the gap between theory and practice as it shows how to integrate the global path planning algorithms in the ROS environment and it proves their efficiency in real scenarios. We believe that this handbook will provide the readers with a comprehensive reference on the global path planning and MRTA problems starting from foundations and modeling, going through simulations and real-world deployments. Links to videos and demonstrations will be included in the book.

Riyadh, Saudi Arabia Anis Koubaa
Riyadh, Saudi Arabia Hachemi Bennaceur
Manouba, Tunisia Imen Chaari
Manouba, Tunisia Sahar Trigui
Riyadh, Saudi Arabia Adel Ammar
Riyadh, Saudi Arabia Mohamed-Foued Sriti
Riyadh, Saudi Arabia Maram Alajlan
Taif, Saudi Arabia Omar Cheikhrouhou
Riyadh, Saudi Arabia Yasir Javed

Acknowledgements

The work is supported by the Robotics and Internet-of-Things (RIOTU) Lab and Research and Initiative Center (RIC) of Prince Sultan University, Saudi Arabia. It also supported by Gaitech Robotics in China. Part of this work was previously supported by the iroboapp project "Design and Analysis of Intelligent Algorithms for Robotic Problems and Applications" under the grant of the National Plan for Sciences, Technology and Innovation (NPSTI), managed by the Science and Technology Unit of Al-Imam Mohamed bin Saud University and by King Abdulaziz Center for Science and Technology (KACST).

Contents

Acronyms

2PPLS	Two-phase Pareto local search
A*	The Astar algorithm
ABC	Artificial bee colony
ACO	Ant Colony Optimization
AD*	Anytime Dynamic A*
AM	Application Master
ANA*	Anytime Nonparametric A*
APF	Artificial Potential Field
ARA*	Anytime Repairing A*
BLE	Broadcast of Local Eligibility
BSP	Bulk Synchronous Parallel
CACO	Conventional ACO
CFor	A set of forbidden configuration
CFree	A set of free configuration
CPD	Compressed path databases technique
CYX	Cycle crossover operator
DWA	Dynamic Window Approach
E*	The E Star algorithm
EDA	Estimation of distribution algorithm
FCE	Free configuration eigen-spaces
FIS	Fuzzy Inference System
FMM	Fast marching method
FOD	Front obstacle distance
GA	Genetic Algorithm
GGA	Grouping genetic algorithms
GGA-SS	Steady-state grouping genetic algorithm
GRASP	Greedy Randomized Adaptive Search Procedure
HACO	Heterogeneous ACO
HDFS	Hadoop Distributed File System
IDPGA	Improved dual-population GA

ILS	Iterated Local Search
IWO	Invasive weed optimization
JPS	Jump point search
LOD	Left obstacle distance
MACO	Modified ACO
MD-MTSP	Multiple Depots MTSP
MLP	Multi-Layer Perceptron
MOKPs	Multi-objective Knapsack problems
MOP	Multi-Objective Optimization
MPCNN	Modified pulsecoupled neural network
MRS	Multi-Robot System
MRTA	Multi-Robot Task Allocation
MTD	Maximum Traveled Distance
MT	Maximum tour
MTSP	Multiple Traveling Salesmen Problem
NM	Node Manager
NN	Neural Networks
ORX	Ordered crossover operator
PFM	The Artificial potential field approach
PFM	Potential field method
PMX	Partially-matched crossover operator
PPaaS	Path Planning as a Service
PRM	The probabilistic roadmap method
PSO	Particle Swarm Optimization
QHS	Quad Harmony Search
RA*	Relaxed AStar
RM	Global Resource Manager
ROD	Right obstacle distance
ROS	Robot Operating System
RRT	Rapidly-exploring random tree
RTMA	Robot and Task Mean Allocation Algorithm
S+T	Services and Tasks
SA	Simulated Annealing
SOM	Self Organizing Maps
SP-CNN	Shortest path cellular neural network
SSSP	Single source shortest path algorithm
TSP	Traveling Salesmen Problem
TS	Tabu Search
TTD	Total Traveled Distance
TWD*	Two Way D*
UAV	Unmanned Air Vehicle
VNS	Variable Neighborhood Search
VRP	Vehicle routing problem
YARN	Yet Another Resource Negotiator

List of Figures

List of Tables

Part I
Global Robot Path Planning

Chapter 1
Introduction to Mobile Robot Path Planning

Abstract Robotic is now gaining a lot of space in our daily life and in several areas in modern industry automation and cyber-physical applications. This requires embedding intelligence into these robots for ensuring (near)-optimal solutions to task execution. Thus, a lot of research problems that pertain to robotic applications have arisen such as planning (path, motion, and mission), task allocation problems, navigation, tracking. In this chapter, we focused on the path planning research problem.

1.1 Introduction

Moving from one place to another is a trivial task, for humans. One decides how to move in a split second. For a robot, such an elementary and basic task is a major challenge. In autonomous robotics, path planning is a central problem in robotics. The typical problem is to find a path for a robot, whether it is a vacuum cleaning robot, a robotic arm, or a magically flying object, from a starting position to a goal position safely. The problem consists in finding a path from a start position to a target position. This problem was addressed in multiple ways in the literature depending on the environment model, the type of robots, the nature of the application, etc.

Safe and effective mobile robot navigation needs an efficient path planning algorithm since the quality of the generated path affects enormously the robotic application. Typically, the minimization of the traveled distance is the principal objective of the navigation process as it influences the other metrics such as the processing time and the energy consumption.

This chapter presents a comprehensive overview on mobile robot global path planning and provides the necessary background on this topic. It describes the different global path planning categories and presents a taxonomy of global path planning problem.

A. Koubaa et al., *Robot Path Planning and Cooperation*, Studies in Computational Intelligence 772, https://doi.org/10.1007/978-3-319-77042-0_1

1.2 Overview of the Robot Path Planning Problem

Nowadays, we are at the cusp of a revolution in robotics. A variety of robotic systems have been developed, and they have shown their effectiveness in performing different kinds of tasks including smart home environments [1], airports [2], shopping malls [3], manufacturing laboratories [4]. An intelligence must be embedded into robot to ensure (near)-optimal execution of the task under consideration and efficiently fulfill the mission. However, embedding intelligence into robotic system imposes the resolution of a huge number of research problems such as navigation which is one of the fundamental problems of mobile robotics systems. To finish successfully the navigation task, a robot must know its position relatively to the position of its goal. Moreover, he has to take into consideration the dangers of the surrounding environment and adjust its actions to maximize the chance to reach the destination.

Putting it simply, to solve the robot navigation problem, we need to find answers to the three following questions: Where am I? Where am I going? How do I get there? These three questions are answered by the three fundamental navigation functions localization, mapping, and motion planning, respectively.

- Localization: It helps the robot to determine its location in the environment. Numerous methods are used for localization such as cameras [5], GPS in outdoor environments [6], ultrasound sensors [7], laser rangefinder [8]. The location can be specified as symbolic reference relative to a local environment (e.g., center of a room), expressed as topological coordinate (e.g., in Room 23) or expressed in absolute coordinate (e.g., latitude, longitude, altitude).
- Mapping: The robot requires a map of its environment in order to identify where he has been moving around so far. The map helps the robot to know the directions and locations. The map can be placed manually into the robot memory (i.e., graph representation, matrix representation) or can be gradually built while the robot discovers the new environment. Mapping is an overlooked topic in robotic navigation.
- Motion planning or path planning: To find a path for the mobile robot, the goal position must be known in advance by the robot, which requires an appropriate addressing scheme that the robot can follow. The addressing scheme serves to indicate to the robot where it will go starting from its starting position. For example, a robot may be requested to go to a certain room in an office environment with simply giving the room number as address. In other scenarios, addresses can be given in absolute or relative coordinates.

Planning is one obvious aspect of navigation that answers the question: ***What is the best way to go there?*** Indeed, for mobile robotic applications, a robot must be able to reach the goal position while avoiding the scattered obstacles in the environment and reducing the path length. There are various issues need to be considered in the path planning of mobile robots due to various purposes and functions of the virtual robot itself as shown in Fig. 1.1. Most of the proposed approaches are focusing on finding the shortest path from the initial position to goal position. Recently, researches are

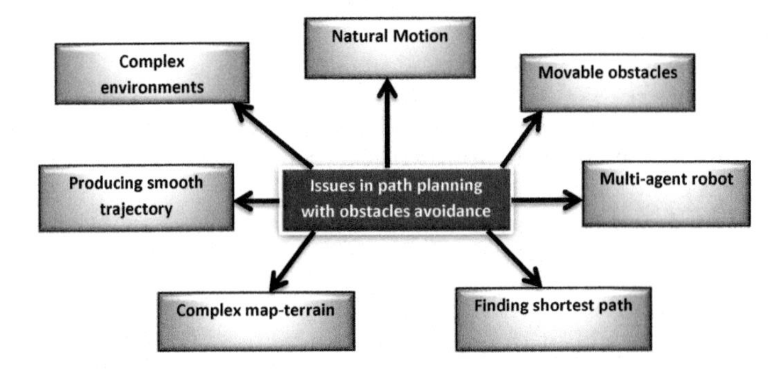

Fig. 1.1 Different issues of path planning

focusing on reducing the computational time and enhancing smooth trajectory of the virtual robot [9]. Other ongoing issues include navigating the autonomous robots in complex environments [10]. Some researchers consider movable obstacles and navigation of the multi-agent robot [11].

Whatever the issue considered in the path planning problem, three major concerns should be considered: efficiency, accuracy, and safety [12]. Any robot should find its path in a short amount of time and while consuming the minimum amount of energy. Besides that, a robot should avoid safely the obstacles that exist in the environment. It must also follow its optimal and obstacle-free route accurately.

Planning a path in large-scale environments is more challenging as the problem becomes more complex and time-consuming which is not convenient for robotic applications in which real-time aspect is needed. In our research work, we concentrate on finding the best approach to solve the path planning for finding shortest path in a minimum amount of time. We also considered that the robot operates in a complex large environment containing several obstacles having arbitrary shape and positions.

1.2.1 Problem Formulation

In [13], Latombe describes the path planning problem as follows:

- $A \subset W$: The robot, it is a single moving rigid object in world W represented in the Euclidean space as $\mathbb{R}^2$ or $\mathbb{R}^3$.
- $O \subset W$: The obstacles are stationary rigid objects in W.
- The geometry, the position, and the orientation of A and O are known a priori.
- The localization of the O in W is accurately known.

Given a start and goal positions of $A \subset W$, plan a path $P \subset W$ denoting the set of position so that $A(p) \bigcap O = \emptyset$ for any position $p \in P$ along the path from start to goal, and terminate and report P or $\emptyset$ if a path has been found or no such path exists. The quality of the path (optimal or not) is measured using a set of optimization

criteria such as shortest length, runtime, or energy.
Four different concepts in this problem need to be well described:

- The moving robot's geometry.
- The workspace (environment) in which the robot moves or acts.
- The degrees of freedom of the robot's motion.
- The initial and the target configuration in the environment, between which a trajectory has to be determined for the robot.

Using this information, we construct the workspace of the mobile robot. Path planning problems can be directly formalized and solved in a 3D workspace. However, these workspace solutions cannot easily handle robots with different geometries and mechanical constraints. To overcome these difficulties, path planning may be formalized and solved in a new space called **the configuration space**. In the configuration space, a robot with a complex geometric shape in a 3D workspace is mapped to a point robot. The robot's trajectory corresponds to a continuous curve in the high-dimensional configuration space, and the environment in which it travels is defined in a two-dimensional plane as depicted in Fig. 1.2. The advantage of the configuration space representation is that it reduces the problem from a rigid body to a point and thus eases the search. The notion of configuration space was first presented by Lozano-Pérez [14]. It contains all the robot configurations. It is denoted by C. In realistic robot applications, the environment in which the robot acts contains obstacles. These obstacles cause some configurations to be forbidden. For example, a configuration c is forbidden if the robot configured at c hits with any of the obstacles in the workspace. More generally, the configuration space C is partitioned into a set of forbidden configurations $C\ for\ b$, and a set of free configurations $C\ free$.

The robot's path is a continuous function $p : [0, L] \rightarrow C$, where L is the length of the path. The path planning problem consists to find an obstacle-free path between a given start configuration $s \in C$ and goal configuration $g \in C$. Formulated in terms of the configuration space C, that is finding a path p such that $p(0) = s$ and $p(L) = g$,

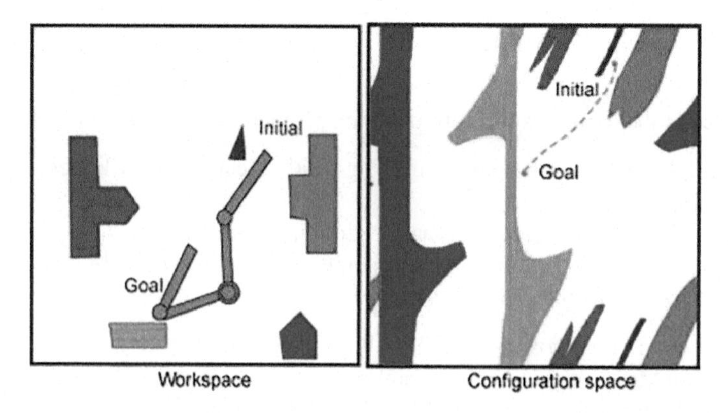

Fig. 1.2 Workspace and configuration space

and $\forall(pos \in [0, L]::p(pos) \in Cfree)$. If such a path does not exist, failure should be reported. In addition to this definition, one may define a quality of measure on all possible paths and require that the path that optimizes this measure is found.

1.3 Path Planning Categories

In this section, we give a classification of the different problems related to mobile robots path planning. It can be divided into three categories according to the robot's knowledge that it has about the environment, the environment nature, and the approach used to solve the problem as depicted in Fig. 1.3.

Environment Nature: The path planning problem can be done in both static and dynamic environments: A static environment is unvarying, the source and destination positions are fixed, and obstacles do not vary locations over time. However, in a dynamic environment, the location of obstacles and goal position may vary during the search process. Typically, path planning in dynamic environments encompasses is more complex than that in static environments due to uncertainty of the environment. As such, the algorithms must adapt to any unexpected change such as the advent of new moving obstacles in the preplanned path or when the target is continuously moving. When both obstacles and targets are moving, the path planning problem becomes even more critical as it must effectively react in real time to both goal and obstacle movements. The path planning approaches applied in static environments are not appropriate for the dynamic problem.

Map knowledge: Mobile robots path planning basically relies on an existing map as a reference to identify initial and goal location and the link between them. The amount of knowledge to the map plays an important role for the design of the path planning algorithm. According to the robot's knowledge about the environment, path planning can be divided into two classes: In the first class, the robot has an a priori knowledge about the environment modeled as a map. This category of path planning

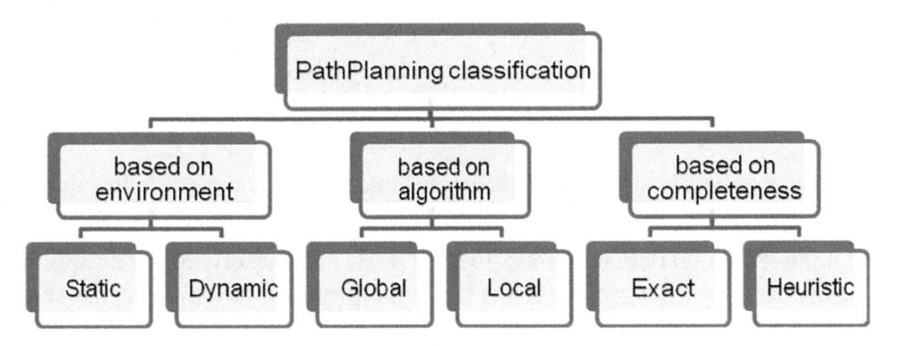

Fig. 1.3 Path Planning Categories

Table 1.1 Global and local path planning

Local path planning	Global path planning
Sensor-based	Map-based
Reactive navigation	Deliberative navigation
Fast response	Relatively slower response
Suppose that the workspace area is incomplete or partially incomplete	The workspace area is known
Generate the path and moving toward target while avoiding obstacles or objects	Generate a feasible path before moving toward the goal position
Done online	Done offline

is known as *global path planning* or *deliberative path planning*. The second class of path planning assumes that the robot does not have a priori information knowledge about its environment (i.e., uncertain environment). Consequently, it has to sense the locations of the obstacles and construct an estimated map of the environment in real time during the search process to avoid obstacles and acquire a suitable path toward the goal state. This type of path planning is known as *local path planning* or *reactive navigation*. Table 1.1 presents the differences between the two categories.

Completeness: Depending on its completeness, the path planning algorithm is classified as either exact or heuristic. An exact algorithm finds an optimal solution if one exists or proves that no feasible solution exists. Heuristic algorithms search for a good-quality solution in a shorter time.

Robot path planning also could be classified according to the number of robots that exist in the same workspace to do the same mission. In many applications, multiple mobile robots cooperate together in the same environment. This problem is called *Multi-Robot Path Planning*. Its objective is to find obstacle-free paths for a team of mobile robots. It enables a group of mobile robots to navigate collaboratively in the same workspace to achieve spatial position goals. It is needed to ensure that any two robots do not collide when following their respective paths.

1.4 Spatial Representations Commonly Used in Path Planning

Roughly, map-based path planning models fall into two categories depending on the way of looking at the world [15].

Qualitative (Route) Path Planning: (Figure 1.4a) In this category, there is no a priori map that represents the world, but it is rather specified by routes from the initial to target location. The world is represented as a connection between landmarks, and a sequence of connected landmarks will represent the route. This approach is similar to how humans describe a route in their natural language (e.g., go straight until reaching

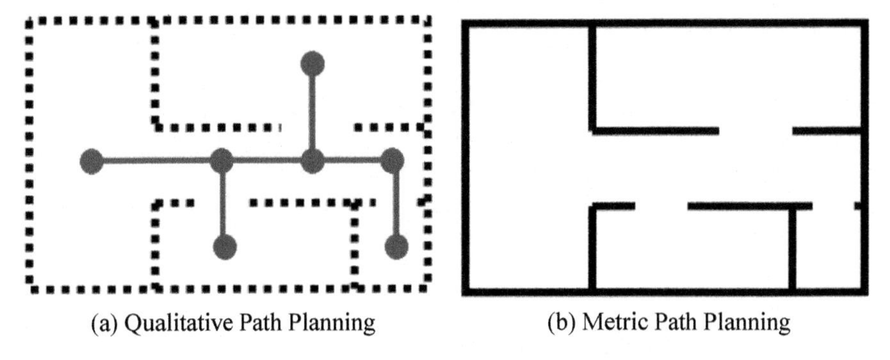

Fig. 1.4 Spatial representations commonly used in path planning

the XYZ restaurant, then turn toward the left, walk until finding the kindergarten, then turn right). Landmarks can be any kind of recognizable and perceivable objects that can uniquely identify a location.

Metric (Layout) Path Planning: (Figure 1.4b) The world is specified using a layout representation that is the map. The map provides an abstract representation to the world (e.g., street layout, intersections, roads). It has to be noted that it is possible to generate route based on layout representation of the environment, but not in other way. In robotics, metric path planning is more attractive than qualitative path planning as it is possible to represent the environment in a clear structure that the robot can use to plan and find its path. Furthermore, it allows the robot to reason about space. In fact, metric maps decompose the environment into a set of distinct points, called waypoints, which are fixed locations identified by their (x,y) coordinates. Planning the path will then consist in finding the sequence of connected waypoints that lead to the goal position while minimizing the path cost. The path cost can be defined as the path length, or path with minimum energy consumption, or the path delay, etc., depending on one of the problem requirements.

1.4.1 Environment Characterization

The robotic environments are characterized in terms of the shape of their obstacles. There are two main types of environments:

Structured environment: in which obstacles have a structured shape and nicely placed within the map. In such a case, the robotic map can easily specify the way the robot can move avoiding obstacles. For instance, in [?], obstacles do not have the actual shape and are represented by dots (for position only) or circles (for position and size). In some other research works [16], environments with only rectangular obstacles are dealt with. In some approaches [17], grids and quad tree are applied to the robot environments for discretization, and obstacles are represented approximately by fitting into the cells.

Unstructured environment or semi-structured: Real robot environments are usually unstructured in which obstacles may have different shape and sizes and be placed in inclined orientation. In such a case, the path planning problem is relatively complex. The algorithm needs to compute a feasible and an optimal path. [18] proposed a real-time path planning of autonomous vehicles for unstructured road navigation.

1.4.2 Path Planning Complexity

If we consider the general formulation of the motion planning problem, where the robot is not simply a point, it has been rigorously proven in this case [19] that the problem is PSPACE-hard (and hence NP-hard). The reason for this is that the configuration space has an unbounded dimension. As for a point robot, the dimension of the working space is bounded (generally 2 or 3), which makes the shortest path problem solvable in polynomial time: ($\mathcal{O}((n)\log(n))$ as an upper bound) [20].

As mentioned in Sect. 1.3, path planning problem could be solved using either exact or heuristic methods. The question is why exponential techniques are used to solve a polynomial problem?

This could be explained by the following reasons:

- Exact methods such as the well-known polynomial Djikstra's algorithm [21] perform a lot of irrelevant computations for reaching the goal. It is not going in the right direction.
- The state space of the problem may be large due to a fine-grain resolution required in the task, or the dimensions of the search problem are high.

It is preferable to use intelligent exact techniques such as Astar (A^*) [22] to guide the search toward promising regions (regions leading to optimal paths) and so avoiding to perform irrelevant computations. On the other hand, the exact methods may only be able to work over low-resolution workspace; if the dimension of the workspace increase, we may observe an increase in computational time.
We can say that in theory the path planning problem as described above has a polynomial complexity, but there is no efficient polynomial technique to solve them in practice due to the above reasons.

1.5 Conclusion

In this chapter, we give a general overview of the path planning problem; the different categories of this problem, approaches used to solve it, its complexity, etc. A lot of intelligent algorithms have been proposed to solve efficiently this problem. These algorithms span over a large number of techniques. This diversity makes hard for a robotic application designer to choose a particular technique for a specific problem. Usually, researchers design and implement a particular algorithm while other techniques might be more effective. This raised the need to devise a classification and taxonomy of the different algorithms being used in the literature. This is the subject of the next chapter.

References

1. Mustafa Al-Khawaldeh, Ibrahim Al-Naimi, Xi Chen, and Philip Moore. 2016. Ubiquitous robotics for knowledge-based auto-configuration system within smart home environment. In *2016 7th international conference on information and communication systems (ICICS)*, pages 139–144. IEEE.
2. Jie-Hua Zhou, Ji-Qiang Zhou, Yong-Sheng Zheng, and Bin Kong. 2016. Research on path planning algorithm of intelligent mowing robot used in large airport lawn. In *2016 international conference on information system and artificial intelligence (ISAI)*, pages 375–379. IEEE.
3. Takayuki Kanda, Masahiro Shiomi, Zenta Miyashita, Hiroshi Ishiguro, and Norihiro Hagita. 2009. An affective guide robot in a shopping mall. In *2009 4th ACM/IEEE international conference on human-robot interaction (HRI)*, pages 173–180. IEEE.
4. Chen, Chiu-Hung, Tung-Kuan Liu, and Jyh-Horng Chou. 2014. A novel crowding genetic algorithm and its applications to manufacturing robots. *IEEE Transactions on Industrial Informatics* 10 (3): 1705–1716.
5. R Visvanathan, SM Mamduh, K Kamarudin, ASA Yeon, A Zakaria, AYM Shakaff, LM Kamarudin, and FSA Saad. 2015. Mobile robot localization system using multiple ceiling mounted cameras. In *2015 IEEE SENSORS*, pages 1–4. IEEE.
6. Takato Saito and Yoji Kuroda. 2013. Mobile robot localization by gps and sequential appearance-based place recognition. In *2013 IEEE/SICE international symposium on system integration (SII)*, pages 25–30. IEEE.
7. Dariush Forouher, Marvin Große Besselmann, and Erik Maehle. 2016. Sensor fusion of depth camera and ultrasound data for obstacle detection and robot navigation. In *2016 14th international conference on control, automation, robotics and vision (ICARCV)*, pages 1–6. IEEE.
8. Luca Baglivo, Nicolas Bellomo, Giordano Miori, Enrico Marcuzzi, Marco Pertile, and Mariolino De Cecco. 2008. An object localization and reaching method for wheeled mobile robots using laser rangefinder. In *2008 4th International IEEE conference intelligent systems, IS'08*, volume 1, pages 5–6. IEEE.
9. Thaker Nayl, Mohammed Q Mohammed, and Saif Q Muhamed. 2017. Obstacles avoidance for an articulated robot using modified smooth path planning. In *2017 international conference on computer and applications (ICCA)*, pages 185–189. IEEE.
10. Ronald Uriol and Antonio Moran. 2017. Mobile robot path planning in complex environments using ant colony optimization algorithm. In *2017 3rd international conference on control, automation and robotics (ICCAR)*, pages 15–21. IEEE.
11. Ram Kishan Dewangan, Anupam Shukla, and W Wilfred Godfrey. 2017. Survey on prioritized multi robot path planning. In *2017 IEEE international conference on smart technologies and management for computing, communication, controls, energy and materials (ICSTM)*, pages 423–428. IEEE.
12. Imen CHAARI DAMMAK. 2012. SmartPATH: A Hybrid ACO-GA Algorithm for Robot Path Planning. Master's thesis, National School of Engineering of Sfax.
13. Jean claude Latombe. 1991. *Robot motion planning*. The Springer International Series in Engineering and Computer Science.

14. Lozano-Pérez, Tomás, A. Michael, and Wesley. 1979. An algorithm for planning collision-free paths among polyhedral obstacles. *Communications of the ACM* 22 (10): 560–570.
15. Murphy, Robin. 2000. *Introduction to AI robotics*. Cambridge, Massachusetts London, England: The MIT Press.
16. Thomas Geisler and Theodore W Manikas. 2002. Autonomous robot navigation system using a novel value encoded genetic algorithm. In *The 2002 45th midwest symposium on circuits and systems, MWSCAS-2002*, volume 3, pages III–III. IEEE.
17. Jianping Tu and Simon X Yang. 2003. Genetic algorithm based path planning for a mobile robot. In *2003 Proceedings of IEEE international conference on robotics and automation, ICRA'03.* , volume 1, pages 1221–1226. IEEE.
18. Chu, K., J. Kim, K. Jo, and M. Sunwoo. 2015. Real-time path planning of autonomous vehicles for unstructured road navigation. *International Journal of Automotive Technology* 16 (4): 653–668.
19. John H Reif. 1985. Complexity of the generalized mover's problem. Technical report, HARVARD UNIV CAMBRIDGE MA AIKEN COMPUTATION LAB.
20. Steven M LaValle. 2006. *Planning algorithms*. Cambridge University Press.
21. Dijkstra, Edsger W. 1959. A note on two problems in connexion with graphs. *Numerische Mathematik* 1 (1): 269–271.
22. Peter, E.Hart, Nils J. Nilsson, and Bertram Raphael. 1968. A formal basis for the heuristic determination of minimum cost paths. *IEEE transactions on Systems Science and Cybernetics* 4 (2): 100–107.

Chapter 2
Background on Artificial Intelligence Algorithms for Global Path Planning

Abstract In the literature, numerous path planning algorithms have been proposed. Although the objective of these algorithms is to find the shortest path between two positions A and B in a particular environment, there are several algorithms based on a diversity of approaches to find a solution to this problem. The complexity of algorithms depends on the underlying techniques and on other external parameters, including the accuracy of the map and the number of obstacles. It is impossible to enumerate all these approaches in this chapter, but we will shed the light on the most used approaches in the literature.

2.1 Introduction

The research on the path planning problem started in late 1960. Nevertheless, most of the efforts are more recent and have been conducted during the 80's [1]. Afterward, several research initiatives, aiming at providing different solutions in both static and dynamic environments, have emerged. Numerous approaches to design these solutions have been attempted, which can be widely classified into three main categories: classical approaches, heuristic approaches, and graph search approaches as depicted in Fig. 2.1. The classical methods are variations and/or combinations of a few general approaches such as Roadmap, potential field, cell decomposition. These methods dominated this field during the first 20 years. However, they were deemed to have some deficiencies in global optimization, time complexity, robustness, etc. The second category of methods used to solve the path planning problem is heuristic approaches which were designed to overcome the aforementioned limits of classical methods. Moreover, numerous numbers of graph search algorithm developed over the last decades have been tested for path planning such as A^*, Dijkstra, Bellman Ford, etc. In this chapter, we will (1) give a description and a classification of the most important approaches of path planning and (2) present an idea about the performance of some important approaches for path planning.

A. Koubaa et al., *Robot Path Planning and Cooperation*, Studies in Computational Intelligence 772, https://doi.org/10.1007/978-3-319-77042-0_2

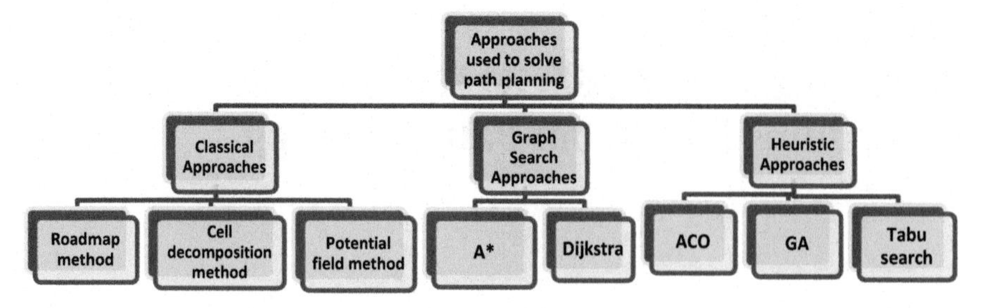

Fig. 2.1 Approaches used to solve the path planning problem

2.2 Classical Approaches

Here, we present the three most successful classical approaches used to solve global robot path planning. For each approach, we present the basic ideas and a few different realizations. For these methods, an explicit representation of the configuration space is assumed to be known.

- *The roadmap approach* and its different methods such as visibility graphs method, freeway method, and silhouette method. The basic idea of roadmap methods is to create a roadmap that reflects the connectivity of Cfree (more details about CFree can be found in Chap. 1). A set of lines, each of which connect two nodes of different polygonal obstacles, lie in the free space and represent a roadmap R. If a continuous path can be found in the free space of R, the initial and goal points are then connected to this path to arrive at the final solution, a free path. If more than one continuous path can be found and the number of nodes on the graph is relatively small, Dijkstra's shortest path algorithm is often used to find the best path. The research works [2–4] use roadmap to solve multi-robot path planning.
- *The cell decomposition approach* and its variants exact cell decomposition, approximate cell decomposition, and wave front planner. The basic idea behind this method is that a path between the initial node and the goal node can be determined by subdividing the free space of the robot's configuration into smaller regions called cells. The decomposition of the free space generates a graph called as a connectivity graph in which each cell is adjacent to other cells. From this connectivity graph, a continuous path can be determined by simply following adjacent free cells from the initial point to the goal point. The first step in cell decomposition is to decompose the free space, which is bounded both externally and internally by polygons, into trapezoidal and triangular cells by simply drawing parallel line segments from each vertex of each interior polygon in the configuration space to the exterior boundary. Then, each cell is numbered and represented as a node in the connectivity graph. Nodes that are adjacent in the configuration space are linked in the connectivity graph. A free path is constructed by connecting the initial configuration to the goal configuration through the midpoints of the intersections of the adjacent cells. The applications of robot path planning based on this method

can be found in [2, 5]. A comparison between the cell decomposition method and the roadmap method is conducted in [2].

- *Artificial potential field approach* (PFM) involves modeling the robot as a particle moving under the influence of a potential field that is determined by the set of obstacles and the target destination. The obstacles and the goal are assigned repulsive and attractive forces, respectively, so that the robot is able to move toward the target while pushing away from obstacles. In [6], the authors implemented the pure PFM method to solve the robot path planning and compared it against an improved PFM and genetic algorithm. In [7], the authors presented an improved PFM to solve path planning in an unknown environment. The authors reinforced the algorithm with a new formula of repelling potential, which is performed in the interest of reducing oscillations and avoiding conflicts when obstacles locate near a target. In [8], Kim proposed a framework based on PFM to escape from a local minimum location of a robot path that may occur under a narrow passage or other possibly similar scenarios.

Limitations: The classical approaches were found to be effective and capable of solving the problem through providing feasible collision-free solutions. However, these approaches suffer from several drawbacks. Classical approaches consume much time to find the solution of the problem, which is considered as a significant drawback especially when dealing with difficult problems of large-scale and complex environments, since they tend to generate computationally expensive solutions. Another drawback of these classical approaches is that they might get trapped in local optimum solutions away from the global optimum solution, especially when dealing with large environments with several possible solutions.

2.3 Graph Search Approaches

Graph search methods are other well-known approaches used to solve the path planning problem. Numerous graph search algorithms developed over the last decades have been tested for path planning of autonomous robots, for instance Astar A^*, Dijkstra, breadth-first search (BFS), depth-first search (DFS). In this chapter, we will focus on Astar (A^*) [9, 10] because it is one of the most efficient algorithms in problem solving. However, this algorithm may be time-consuming to reach the optimal solution for hard instances depending on the density of obstacles [11–13].

2.3.1 The AStar (A) Algorithm*

2.3.1.1 Overview

The A^* (AStar) algorithm is a path finding algorithm introduced in [9]. It is an extension of Dijkstra's algorithm [14]. A^* achieves better performance (with respect to time), as compared to Dijkstra, by using heuristics. The A^* algorithm is presented in Algorithm 1.

In the process of searching the shortest path, each cell in the grid is evaluated according to an evaluation function given by:

$$f(n) = h(n) + g(n) \tag{2.1}$$

where $g(n)$ is the accumulated cost of reaching the current cell n from the start position S:

$$g(n) = \left\{ \begin{array}{r} g(S) = 0 \\ g(parent(n)) + dist(parent(n), n) \end{array} \right\} \tag{2.2}$$

$h(n)$ is an estimate of the cost of the least path to reach the goal position G from the current cell n. The estimated cost is called heuristic. $h(n)$ can be defined as the Euclidian distance from n to G. $f(n)$ is the estimation of the minimum cost among all paths from the start cell S to the goal cell G. The Tie-breaking factor $tBreak$ multiplies the heuristic value ($tBreak * h(n)$). When it is used, the algorithm favors a certain direction in case of ties. If we do not use tie-breaking, the algorithm would explore all equally likely paths at the same time, which can be very costly, especially when dealing with a grid environment. In a grid environment, the tie-breaking coefficient can be chosen as:

$$tBreak = 1 + 1/(length(Grid) + width(Grid)) \tag{2.3}$$

The algorithm relies on two lists: the open list which is a kind of a shopping list. It contains cells that might fall along the best path, but may be not. Basically, this is a list of cells that need to be checked out.

The closed list: It contains the cells already explored.

Each cell saved in the list is characterized by five attributes: ID, $parentCell$, g_Cost, h_Cost, and f_Cost.

The search begins by expanding the neighbor cells of the start position S. The neighbor cell with the lowest f_Cost is selected from the open list, expanded and added to the closed list. In each iteration, this process is repeated. Some conditions should be verified while exploring the neighbor cells of the current cell, and a neighbor cell is:

1. Ignored if it already exists in the closed list.
2. If it already exists in the open list, we should compare the g_cost of this path to the neighbor cell and the g_cost of the old path to the neighbor cell. If the g_cost of using the current cell to get to the neighbor cell is the lower cost, we change the parent cell of the neighbor cell to the current cell and recalculate g, h, and f costs of the neighbor cell.

This process is repeated until the goal position is reached. Working backward from the goal cell, we go from each cell to its parent cell until we reach the starting cell (the shortest path in the grid map is found).

Algorithm 1. The standard Astar Algorithm A^*

```
   input : Grid, Start, Goal
   // Initialisation:
1  closedSet = empty set // Set of already evaluated nodes;
2  openSet = Start // Set of nodes to be evaluated;
3  came_from = the empty map // map of navigated nodes;
4  tBreak = 1+1/(length(Grid)+width(Grid)); // coefficient for
   breaking ties;
5  g_score[Start] = 0; // Cost from Start along best known
   path;
   // Estimated total cost from Start to Goal:
6  f_score[Start] = heuristic_cost(Start, Goal);
7  while openSet is not empty do
8      current = the node in openSet having the lowest f_score;
9      if current = Goal then
10         return reconstruct_path(came_from, Goal);
11     end
12     remove current from openSet;
13     add current to closedSet;
14     for each free neighbor v of current do
15         if v in closedSet then
16             continue;
17         end
18         tentative_g_score = g_score[current] + dist_edge(current, v);
19         if v not in openSet or tentative_g_score < g_score[v] then
20             came_from[v] = current;
21             g_score[v] = tentative_g_score;
22             f_score[v] = g_score[v] + tBreak * heuristic_cost(v, Goal);
23             if neighbor not in openSet then
24                 add neighbor to openSet;
25             end
26         end
27     end
28 end
29 return failure;
```

2.3.1.2 A^* for Robot Path Planning: Literature Review

Several modified A^* versions were proposed to solve the path planning problem. The main idea used to build such variants of A* is the relaxing of the admissibility criterion ensuring an optimal solution path. These relaxations are often bounded by weighting the admissible heuristic in order to guarantee near-optimal solutions no worse than $(1+\epsilon)$ times the optimal solution path. This heuristic is referred to as ϵ-admissible. For instance, the relaxation Weighted A* [15] uses the ϵ-admissible heuristic $hw(n) = \epsilon * h(n)$, and $\epsilon \geq 1$ is shown to be faster than the classical A* using the admissible heuristic h since fewer nodes are expanded. Moreover, in the worst case, the gap between the path solution found by that algorithm and the optimal solution is of order c*$(\epsilon - 1)$ where c* is the cost of the optimal solution. The static weighting relaxation [16] uses the cost function $f(n) = g(n) + (1+\epsilon) * h(n)$. The dynamic weighting [17] uses the cost function $f(n) = g(n) + (1+\epsilon * w(n)) * h(n)$, where $w(n)$ is a function of the depth of the search. Sampled dynamic weighting [18] uses sampling of nodes to better estimate the heuristic error. Many other relaxed A* approaches were proposed, most of them relax the admissibility criterion. In [19], the authors proposed an intelligent exploration of the search space, called jump points, by selectively expanding only certain nodes in a grid map. They defined a strategy to prune a set of nodes between two jump points so that intermediate nodes on a path connecting two jump points are never expanded. They showed that this approach is optimal and can significantly speed up A*. Also to enhance the performance of A*, different optimizations were proposed in [20]. The optimizations are related to the data structure used to implement A*, the heuristics, and the re-expansion of nodes. Indeed, the authors proved that the use of an array of stacks instead of a list to implement the open and the closed lists speeds up the A* algorithm. They also tested different heuristics including the Manhattan heuristic, the ALT heuristic, and the ALTBestp heuristic, and they showed that ALTBestp heuristic is better than the two other heuristics for both A* and IDA* [21] with an array of stacks. To avoid re-expanding nodes already expanded by going through the open and the closed lists, a two-step lazy cache optimization strategy has been proposed. In [22], the authors present a performance evaluation of TRANSIT routing technique. TRANSIT is a technique for computing shortest path in road networks. They tested it on a set of grid-based video-game benchmarks. They claimed that TRANSIT is strongly and negatively impacted, in terms of running time and memory, by uniform-cost path symmetries. To solve this problem, the authors designed a new technique for breaking symmetries using small additive epsilon costs to perturb the weights of edges in the search graph. Simulation results showed that the modified TRANSIT technique reduces the execution time as compared to A* and the compressed path database (CPD) technique [23]. It also reduces the number of expanded nodes. Likhachev et al. [24] proposed the Anytime Repairing A* (ARA*) algorithm for path planning. ARA* runs a series of A* with weighted heuristics, but it reuses information from previous executions. ARA* uses a weighting value (ϵ) (like explained above) to rapidly find an initial suboptimal path and then reduces ϵ to improve path quality over time. ARA* finds an initial solution for a given initial value of ϵ and continues the search with progressively smaller values of ϵ to improve the solution and reduce its suboptimality bound. The value of ϵ is decreased by a fixed amount each time an improved solution is found, and the current-best solution is proven to be ϵ-suboptimal. The f(s)-values of the states s $\in$ OPEN list are updated to account for the new value of ϵ. The initial value of ϵ and the amount by which it is decreased in each iteration are parameters of the algorithm. The same team later presented the AD* algorithm (An Anytime, Re-planning Algorithm) [25], an algorithm that combines the benefits of anytime and re-planning algorithms to solve path planning in dynamic, relatively high-dimensional state spaces. The AD* combines the D* lite [26] and the ARA* algorithm [24]. The core idea of the algorithm consists in generating a set of solutions with improved bound after executing a series of searches while decreasing the inflation factors. When there are changes in the environment affecting the cost of edges in the graph, locally affected states are placed on the OPEN queue with priorities equal to the minimum of their previous key value and their new key value, as with D* Lite. States on the queue are then processed. To evaluate the performance of AD*, they compared it against ARA* and D* lite on real-world robotic application, and they proved that the novel technique is very efficient. In [27], a new Anytime A* algorithm called Anytime Nonparametric A* (ANA*) is proposed. The algorithm improves on ARA* in five ways: (1) ANA* does not require parameters to be set; (2) ANA* is maximally greedy to find an initial solution; (3) ANA* is maximally greedy to improve the current-best solution; (4) ANA* gradually decreases the suboptimality-bound current-best solution; and (5) ANA* only needs to update the keys of the states in the OPEN queue when an improved solution is found.

2.4 Heuristic Approaches

Classical methods dominated the field of robot path planning before 2000 but had much lower percentage of usage after that year. Another category of approaches has been found more popular in the robot navigation field compared to classical techniques as shown in Fig. 2.2. This category was designed to overcome the aforementioned limit of the classical and exact methods, and they are called metaheuristics. The term metaheuristic was introduced in first time by [28], which originates from the composition of two Greek words. Heuristic derives from the verb heuriskein which means to find, while meta means beyond. Osman and Laporte [29] defines the metaheuristics "as an iterative generation process which guides a subordinate heuristic by combining intelligently different concepts for exploring and exploiting the search space, learning strategies are used to structure information in order to find efficiently near-optimal solutions," while [30] defines metaheuristic as an iterative master process that guides and modifies the operations of subordinate heuristics to efficiently produce high-quality solutions. It may manipulate a complete (or incomplete) single solution or a collection of solutions at each iteration. The subordinate heuristics may be high (or low)-level procedures, or a simple local search, or just a construction method. This category of algorithms includes but is not limited to ant

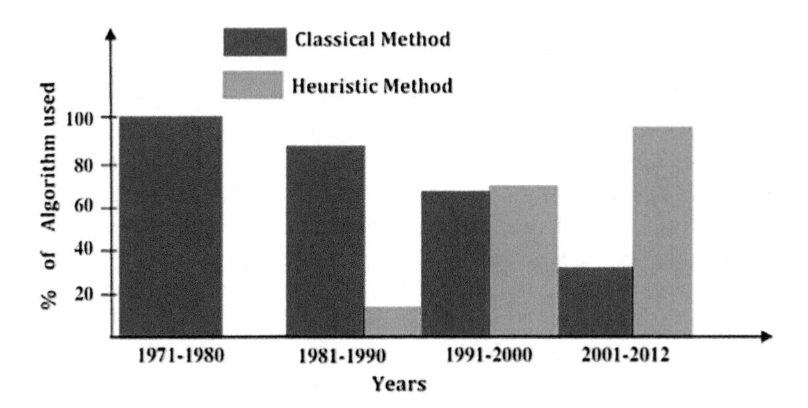

Fig. 2.2 Application of classical and heuristic algorithms [31]

colony optimization (ACO), particle swarm optimization (PSO), genetic algorithm (GA), iterated local search (ILS), simulated annealing (SA), Tabu Search (TS), neural network (NN), The greedy randomized adaptive search procedure (GRASP), and variable neighborhood search (VNS). One of the main properties of the metaheuristic approaches is the use of probabilistic decisions during the search. Both metaheuristic algorithms and pure random search algorithms are characterized by the randomness aspect. Metaheuristics uses it in an intelligent manner while the pure random search uses it in a blind manner.

Characteristics of metaheuristics:

- Metaheuristic algorithms are approximate and usually non-deterministic.
- They may incorporate mechanisms to avoid getting trapped in local optimum.
- Metaheuristics are not problem-specific.
- Metaheuristics may make use of domain-specific knowledge about the form of heuristics that are controlled by the upper level strategy.

In this section, we will present some metaheuristics considered in our research work. The study covers four metaheuristics: neural network (NN), some of the most common nature-inspired algorithms which are GA and ACO and local search methods such as Tabu Search. In the beginning of each subsection, an introduction of the basic approach is provided and then some related works are presented.

2.4.1 Tabu Search

2.4.1.1 Overview

To deal with combinatorial problems, which are increasingly difficult to solve with a computer, local search methods were used in the 70s whose principle is very simple and manage to solve some hard combinatorial problems for which exhaustive search methods were powerless. Starting from some given solution, called current solution, which is generated randomly or computed using a greedy algorithm, classical local search algorithms iteratively try to improve the current solution around an appropriate defined neighborhood. When a better solution is found, the local search continues the search from it. When no better solution can be found in the neighborhood of the current solution, the local search stops in a local optimum. The drawback of local search algorithm is that it may get stuck at a poor quality local optimum. To overcome this drawback, many metaheuristics were proposed. In 1983, a new metaheuristic appears called simulated annealing [32]. It allows random exploration of the search space for escaping local optimums. In 1986, a new metaheuristic called Tabu Search (TS) was introduced by Fred Glover [33], and he provided a full description of the method in 1989 [34]. The basic principle of TS metaheuristic is to continue the search for solutions even when a local optimum is encountered by temporary accepting moves degrading the current solution in order to escape local optimums.

TS has been used to solve a wide range of hard optimization problems such as job shop scheduling, graph coloring, the traveling salesman problem (TSP), and the quadratic assignment problem.

2.4.1.2 Tabu Search Concepts

In what follows, we describe the different concepts of the Tabu Search algorithm. The basic schema of TS algorithm is presented in Algorithm 2.

TS simulates the problem-solving process by taking advantage of the search history. Indeed, TS is a variable neighborhood method, which employs "intelligent" search and flexible memory techniques to avoid being stuck at a local optimum and revisiting previously seen solutions. As a local search algorithm, the Tabu Search method tries, in each iteration, to ameliorate the current solution around its neighborhood until a fixed stopping condition is satisfied. In order to create the different neighbor solutions, a small transformation called *move* is applied to the current solution. The move may result in either the improvement of the current solution or its deterioration. Exploring completely the neighborhood of the current solution may be too time-consuming. To overcome this problem, the following solutions have been suggested: (1) *Best among all.* It checks all the current solution neighborhoods and chooses the best neighboring solution that has the best cost. This method is time-consuming as it generates and evaluates all the candidates in the neighborhood. (2) *First-improving*. The first solution found in the neighborhood that has a cost smaller than the current solution is chosen. This method has the advantage of being fast, as the search of neighborhood is stopped once a new better solution is found. (3) *Best among first C (C random).* This method consists to generate a set of C neighboring solution where C is random number. Then, the best solution (that has the lower cost than the current solution) is selected. (4) *Best among first C (C constant).* First, C solutions of the current solution are created where C is constant. The solution which has a lower cost than the current solution is selected.

To memorize the visited regions of the state space of the problem, Tabu Search put in place a mechanism avoiding the search to rapidly returning to these regions called *Tabu List*. The Tabu List is a kind of memory containing temporary tabu moves as attributes. A move is *tabu* when it is not permitted to be executed. A tabu move remains on the list during some iterations of the search. The size of the Tabu list is referred to as *tabu tenure*. The duration of keeping a tabu move within the list depends on some parameters of the search in order to guarantee the absence of cycles. For a given problem, it may exist many ways to design and specify the tabu list mechanism. The tabu list mechanism is described by specifying its structure and its size, how it is updated after carrying out a move, which moves or neighborhoods are not allowed according to the content of the list, and how long (number of iterations) a neighborhood remains tabu. The behavior of the TS strongly depends on the tabu tenure, if the tabu list is too short, there are a few tabu moves, then the probability to meet a cycle increases and the algorithm explores only a few regions of the search space. On the other hand, if the tabu list is too long, there are more tabu moves, this

Algorithm 2. The Tabu Search Algorithm

```
1  Generate an initial solution s; // random or greedy algorithm may
       be used
2  s*=s; // s* is the best current solution
3  TL= ; // tabu list initially empty
4  while a stop condition is not reached do
5  |   M ← the set of all non tabu moves plus some exceptional tabu moves
   |   (aspiration criteria) ;
6  |   m ← best move of M ;
7  |   s ← m(s); // perform move m on s to get a new
   |       configuration
8  |   update TL ;
9  |   if f(s*) > f(s) the s* ← s; // improve the current best
   |       solution
10 end
11 return s* ;
```

allows to explore further the search space (many regions of the search space will be covered), and the probability to meet a cycle is reduced. However, the search is likely to miss several local optima. A good TS algorithm must make a good balance in terms of memory by fixing an appropriate value for the tabu tenure. In order to study and improve the performance of TS, several ways of adjusting the size of the tabu list, expressed in terms of the number of iterations, were proposed as:

- Simply keep it constant during all the search.
- Randomly selecting it between two values min and max.

Dynamically varying it during the search; if the search tends to cycle or to stagnate, the size of the tabu list is increased; otherwise, it is slightly reduced. In certain cases, some tabu moves may be judged useful for improving the search. So, it is suitable to liberate them from the tabu list; then, according certain conditions called aspiration criteria, some tabu moves can be allowed during the search. The aspiration criterion is a method used to relax the mechanism of the tabu list. The aspiration mechanism determines a criterion for which a tabu move would be accepted. For instance, a basic aspiration criterion would be to allow tabu moves improving the current-best solution. However, one must ensure that this criterion will not lose the main role of the tabu list, in particular the presence of cycles during the search. The stopping conditions in Tabu Search algorithm may be:

- The allowed maximum number of iterations or max time is reached.
- A satisfactory solution was found; a lower bound of the problem can be used to evaluate the solution quality.
- A known optimal solution is obtained.
- The search is stagnated (the best current solution was not improved along a certain number of iterations).

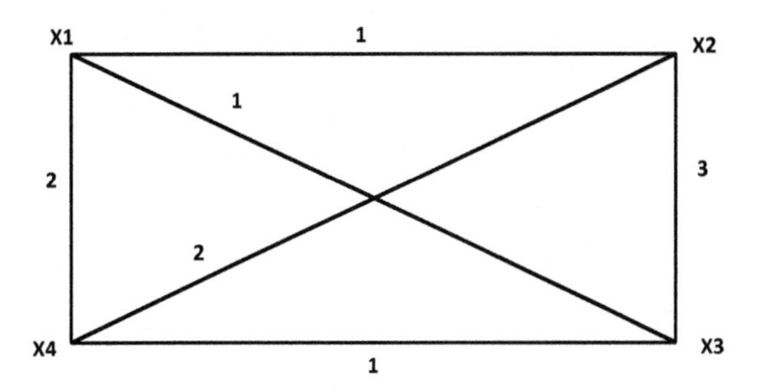

Fig. 2.3 Simple illustrative example of the Tabu Search algorithm

2.4.1.3 Illustrative Example

To illustrate the different concepts used in TS, let us consider the following simple path planning problem (Fig. 2.3) modeled as a traveling salesman problem which consists to find an optimal tour passing once through every vertex.

A feasible solution is a tour passing once through every vertex, for instance (1, 2, 3, 4, 1) with cost 7. The state space consists of all feasible solutions, and there are $(n-1)! = 6$ feasible solutions. The $(n-1)!$ feasible solutions should be examined in order to determine an optimal solution. Hence, we find that an optimal solution is the tour (1, 3, 4, 2, 1) with cost=5.

There are many ways to encode and define a move for this problem. In our example, a feasible solution is encoded by the set of vertices of the tour, and a move consists in swapping any two vertices to get a new feasible solution. For instance, the neighborhood of the solution (1, 2, 3, 4, 1) = {(1, 3, 2, 4, 1), (1, 4, 3, 2, 1), (1, 2, 4, 3, 1)}. The swap is shown in blue color.

The Tabu list may be implemented as a matrix memorizing tabu moves as follows: Each cell represents the tabu tenure of a swap operation. If two vertices are swapped, for example, vertex 1 and vertex 2, the move will be made tabu in the matrix in both [1][2] and [2][1]. This is to prevent the algorithm from going back on a move before exploring a little.

2.4.1.4 Improvements of the Tabu Search Strategy

The basic Tabu Search is improved by using two important operations:

1. **Intensification**: Intensify searching in neighborhoods that seem to have a chance to lead to solutions close to the optimum. This operation aims to make sure that best solutions in these regions are indeed found. Based on a ***recency memory*** (short-term memory), it records the number of consecutive iterations that various solution components have been present in the current solution without interruption. Two approaches may be used to implement this operation:

- Restart the search from the best currently known solution by fixing the components that seem to be more attractive.
- Restart the search from the best currently known solution by modifying the neighborhood (increasing the size of the neighborhood) in such way to allow more moves.

2. **Diversification**: After a certain number of consecutive iterations, the search processus is stagnated. The diversification mechanism can be used to move the search toward a new area. If the search is too local, then it is displaced from the current neighborhood to another one. High priority is given to solutions of another neighborhood than the current one. Moves that are not applied or rarely applied are tried. Based on a ***frequency memory*** (long-term memory), it records the number of total iterations from the beginning that various solution components have been present in the current solution. This operation can be implemented by restarting the search from a solution that contains components rarely used.

2.4.1.5 Robot Path Planning Using Tabu Search: Literature Review

Tabu Search approach has been proved to be a very successful metaheuristic in dealing with combinatorial problems. However, in the literature, few research works based on Tabu Search have been proposed to solve the path planning problem. For example, [35, 36] research works were the first, as claimed in those papers, to present a solution to the local path planning problem using the TS approach where they demonstrated its effectiveness. The idea of the algorithm is to find, in each iteration, a set of tabu moves to confine the robot's locations and to guide its motion toward the destination position. The authors also proposed an effective cost function to build a directed path and avoid local minima. The effectiveness of the approach was demonstrated through simulations. The time complexity of the TS-based planner was reported to be O(n), and a solution is guaranteed to be found if a valid path exists. We note that this approach was applied for a static environment modeled as a grid map. In [37], a comparative study was conducted between three different approaches classified as population-based approaches and trajectory based for solving the global path planning problem of mobile robots: Tabu Search, simulated annealing, and genetic algorithms. For the Tabu Search algorithm, the neighboring solutions are generated by selecting two non-consecutive nodes from the current solution, then generating a new random subpath between the two previously selected nodes, and concatenating the subpath with the remainder of the current path. The negative point of this work is that it did not give any information about the solution length (number of nodes included in a solution), the neighborhood size, and the method followed to evaluate the costs of neighbors, whereas these parameters impact enormously on the effectiveness of this method. In addition, the planners were not compared against any exact method. Simulation was performed on one example of the campus map. It was proved that simulated annealing planner performs better than the other planners in terms of runtime, while Tabu Search was demonstrated to offer the best solution quality.

In [38], the authors presented a new algorithm to solve the local path planning problem (sensor-based path planning). They proposed a sampling-based path planner that incorporates a Tabu Search procedure to make the robot motion more intelligent by restricting and guiding the sampling area to the more promising regions of the free space. In the conventional sampling method, the next position of the robot is chosen randomly; in this paper, the authors integrate a Tabu Search component to select the best next position of the robot. Two strategies are used in the sampling method: uniform sampling and Gaussian sampling. The Tabu Search procedure has three phases: local search, intensification, and diversification. In the local search, two tabu lists are used: short tabu list and long tabu list to search the best next position of the robot. The short tabu list obliges the robot to move to the next position which is far away from the current position and close to the goal position. The long tabu list records all the positions already visited by the robot to prevent revisiting them again. The intensification and diversification are used when the sampling method is not able to find a next position for the robot. The authors tested the algorithm in three different classes of environments: convex, concave, and maze environments, they also compare the algorithm with different other approaches, and they proved its performance in terms of path cost and execution time. In [39], the authors introduced an algorithm for robot path planning (SLEEP) based on Tabu Search approach, and they consider in their implementation both energy consumption and staying alive for rechargeable robots. The environment model was represented by a directed graph. Each node of the graph represents a task that the robot must accomplish. The problem is a variation of the traveling salesman problem (TSP) as the robot must return to the initial node for battery changing or recharging. The authors introduced first an angle-based traveling algorithm that can generate a good initial path planning. They presented then a TS method which is used to obtain the best order of tasks while minimizing the energy consumption. To generate the neighborhood of the current solution, two moves are used "swap and insert." Experiments are conducted on a real mobile robot, "UP-Voyager IIA". The algorithm is compared to other approaches that exist in the literature. The authors tested the algorithm with different number of tasks, and they evaluate the energy consumption of the robot. They proved that SLEEP technique can provide an effective path planning by which the robot can be guaranteed to stay alive and finish all tasks with the minimum energy consumption. Wang et al. in [40] proposed an approach based on Tabu Search to solve the staying-alive and energy-efficient path planning problem. The problem is a variation of the traveling salesman problem. The robot has to finish a set of tasks efficiently while staying alive during the execution. The authors presented in their method the initial solution generation, the tabu candidate selection strategies, the objective function with penalty factor for staying-alive and tabu properties to solve the problem. They demonstrated that the TS approach can provide an effective path planning by which a robot can be guaranteed to stay alive and finish all tasks with the minimum energy as compared to the greedy TSP method.

2.4.2 *Genetic Algorithms*

2.4.2.1 Overview

The genetic algorithms (GA) is a metaheuristic created by John Holland in 1975 [41]. It simulates the process of genetic evolution and the natural selection to solve optimization and research problems. GA is a subset of a much larger branch of computation known as evolutionary computation.

2.4.2.2 Genetic Algorithm's Basic Concepts

Before beginning a discussion on genetic algorithms, it is essential to be familiar with some basic terminology:

- Population: It is a subset of all the possible (encoded) solutions to the given problem.
- Chromosomes: A chromosome is one such solution to the given problem. A set of chromosomes form a population.
- Gene: A gene is one element position of a chromosome.
- Fitness Function: A fitness function simply defined is a function which takes the solution as input and produces the suitability of the solution as the output.
- Genetic Operators: These operators alter the genetic composition of the chromosomes during the execution of the algorithm.

In GA, we have a pool or a population of possible solutions to the given problem, and the initial population may be generated at random or seeded by other heuristics. A population is characterized by its size, the size should not be kept very large as it can cause a GA to slow down, while a smaller population might lead to premature convergence. Therefore, an optimal population size needs to be decided by trial and error.

The evolution from one population to another population invokes four steps during which the chromosomes undergo different genetic operators: (1) fitness evaluation, (2) selection, (3) crossover, and (4) mutation (like in natural genetics): producing new children called offspring, and the process is repeated over various generations. Each individual (or candidate solution) is assigned a fitness value (based on its objective function value), and the fitter individuals are given a higher chance to mate and yield "fitter" individuals. This process is repeated until a termination condition is met. This is in line with the Darwinian theory of "Survival of the Fittest." In this way, the algorithm keeps "evolving" better individuals or solutions over generations till it reaches a stopping criterion.

Genetic algorithms are sufficiently randomized in nature, but they perform much better than random local search, as they exploit historical information as well. One of the most important decisions to make while implementing a genetic algorithm is how to represent the solutions. It has been observed that improper representation

Algorithm 3. The Genetic Algorithm

```
1  Generate the initial population;
2  while (Number of generation < Maximum number of generation) do
3  |   Perform fitness evaluation for each individual;
4  |   Perform the elitist selection;
5  |   Perform the Rank selection;
6  |   repeat
7  |   |   From current generation choose two parents randomly;
8  |   |   if (random number generated < the crossover probability) then
9  |   |   |   perform the crossover;
10 |   |   |   move the resultant chromosomes to the next generation;
11 |   |   else
12 |   |   |   move the parents to the next generation;
13 |   |   end
14 |   until (Population size of next generation = Maximum size of population);
15 |   for each individual in the next generation do
16 |   |   if (random number generated < the mutation probability) then
17 |   |   |   perform mutation;
18 |   |   end
19 |   end
20 end
```

can lead to poor performance of the GA. There are several representations used in the literature; for instance, the binary representation is one of the simplest and most widely used representations in GA. In this type of representation, the chromosome is composed from a set of bits "1" and "0." Another type of representation is integer representation in which the chromosome is represented by a set of integers.

Parent selection is the process of selecting parents which mate and recombine to create offsprings for the next generation. Parent selection is very crucial to the convergence of the GA as good parents drive individuals to a better and fitter solutions. Different selection methods have been defined in the literature such as roulette wheel selection, tournament selection, rank selection.

The crossover operator is analogous to reproduction and biological crossover. In this, more than one parent is selected and one or more offsprings are produced using the genetic material of the parents. Crossover is usually applied in a GA with a high probability p_c. The most popularly used crossover operators are one-point crossover, two-point crossover, and uniform crossover.

Mutation may be defined as a small random tweak in the chromosome, to get a new solution. It is used to maintain and introduce diversity in the genetic population and is usually applied with a low probability p_m.

2.4.2.3 Genetic Algorithm's Pseudocode

The basic schema of GA is presented in Algorithm 3

2.4.2.4 Stopping Condition

The termination condition of a genetic algorithm is important in determining when a GA run will end. It has been observed that initially the GA progresses very fast with better solutions coming in every few iterations, but this tends to saturate in the later stages where the improvements are very small. We usually want a termination condition such that our solution is close to the optimal, at the end of the run.

Usually, we choose one of the following termination conditions:

- When there has been no improvement in the population for X iterations.
- When we reach an absolute number of generations.
- When the objective function value has reached a certain predefined value.

2.4.2.5 Illustrative Example

To illustrate the different concepts used in GA, let us consider the following simple path planning problem for 5*5 grid map. The white square is free cells in the map, and the black square is the obstacles that should be avoided (Fig. 2.4). The problem consists to find a path between the start position 0 and the goal position 24. Let us consider that the maximum number of generations is 3 and the population size is 4. We will present only the different steps of the first iteration. The algorithm stops when the number of iterations reaches the maximum number of iterations.

Iteration0: The first step consists to generate the initial population and evaluate the cost (fitness value) of each path. The set of paths generated are:
0-6-7-8-9-14-19-24: cost=7.4
0-6-7-8-14-19-24: cost=6.8
0-5-10-11-6-7-13-19-24: cost=8.8
0-5-11-7-13-19-24: cost=7.6
The second step consists to make the crossover operator on the selected parents.
parent1: 0-6-7-8-9-14-19-24 cost=7.4
Parent2: 0-5-11-7-13-19-24 cost=7.6
In this example, we will perform one-point crossover. The commonly chosen cell is 7, and thus, the following two offsprings are generated.
Offspring1: 0-6-7-13-19-24: cost 6.2
Offspring2: 0-5-11-7-8-9-14-19-24: cost=10.2

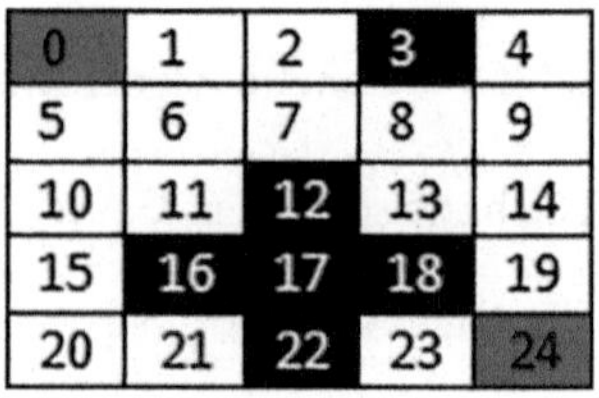

Fig. 2.4 5*5 grid map

0	1	2	3	4
5	6	7	8	9
10	11	12	13	14
15	16	17	18	19
20	21	22	23	24

The next step is the mutation on a randomly chosen path from the new generation 0-5-11-7-8-9-14-19-24: cost = 10.2, and the second cell in the path will be changed by the neighbor cell 6. Thus, we obtain the following path: 0-6-11-7-8-9-14-19-24: cost = 8.8

2.4.2.6 Robot Path Planning Using Genetic Algorithms: Literature Review

From enhancing the initial population to proposing other approaches to improve the GA performance, a lot of research efforts were conducted. In what follows, we present some relevant works that presented solutions for the robot path planning problem using GA.

1. **Enhancing GA Initial Population**: In [42], the authors proposed enhancing and reducing the initial population size by using rough set reduction theory. In [43], authors proposed using a probability-based fast random search method to generate the initial individuals with higher advantage. The work in [44] proposed a position information feedback-based method generated using ACO and a priority grouping method to generate an initial population containing only feasible paths. Their simulation results show that the proposed approach provides better solutions in a higher speed.
2. **Variation of GA Basic Characteristics**: A GA based on visibility graphs was proposed in [45]. New operator aiming at enhancing the performance was added to the GA. The author opted to use fixed length chromosomes to represent the solution. Each chromosome consists of string of bits which represent the vertices of the graph. Comparing the GA with SA and hill-climbing methods, the results indicate that the GA shows better performance in complex environment and it was capable of finding near-optimal solutions. In [46], authors proposed a fast GA-based approach by improving the evaluation function and the natural selection process. The proposed method increases the number of feasible subpaths in the early stages and minimizes the distances of the paths in the later stages. The proposed method shows its effectiveness comparing with traditional GA and two other previously proposed GA-based approaches in four 16*16 environments. In [47], Sedighi et al. proposed a GA approach using new chromosome structure based on variable-monotone approach. Simulations were conducted on ten different search spaces with varying obstacle complexity. The results show that the proposed approach significantly outperformed the previous single-monotone approaches in complex cases, i.e., high obstacles ratio. In [48], authors proposed a new adaptive approach based on the GA to solve the robot path planning problem. A novel selection operator was designed to escape the local optimum problem and premature convergence. At each iteration, if necessary, the selective pressure is updated by using feedback information from the standard deviation of fitness function values. The proposed approach outperforms three previous GA-based approaches in terms of solution quality.

In [49], Liu et al. presented knowledge-based genetic operators to improve the GA performance. They added active constraint to create the initial population, and some evaluation was done before performing the crossover operator. Multipoint crossover was used with variable crossover probability. The mutation is used to add genes between any two disconnected parts in the paths. Variable length chromosomes are adopted where each gene is represented by two-dimensional coordinates. The performance of the proposed approach was tested using simulations for single and multi (two)-robots in three different environments. The results show that the GA with knowledge-based operators can find a near-optimal solution and satisfies real-time demand. The work in [50] proposed guiding the GA search with additional knowledge about the problem by developing new optimization operators. These operators are used to improve the quality of the solution by replacing or removing some nodes. The genetic operators in the proposed approach may produce infeasible paths, and they are not discarded from the search but penalized with some additional cost using modified fitness function. Furthermore, in [51], the authors proposed a modified GA incorporating with problem-specific knowledge. After performing the genetic operators, the resultant paths are subjected to one further optimization process. The proposed approach was adapted for both static and dynamic environments. Comparing with the traditional GA, the proposed approach can find its optimal solution with fewer number of generations and it is more efficient in dynamic environment. The GA was also adapted to solve multi-objective optimization problems. The work in [52] proposed a modified GA with problem-specific knowledge operators to solve the path planning problem with multi-objective optimization in complex and static environment. Three objectives were considered to be optimized: traveling distance, path smoothness, and path safety. A modified A* algorithm was used to generate the initial population for the GA in complex cases, i.e., high obstacles ratio. Also, new operators were proposed to help avoiding the local optimum problem. First, operator is a deletion operator, which is used to remove any useless moves such as loops in the path. Second operator is an enhanced mutation operator, which is used to improve the paths partially by selecting two points and reconnect them using basic A* algorithm. This work was extended in [53], by considering more objectives to be optimized: traveling time and energy consumption. Also, an approach to convert the optimal path in complex cases to a time-based profile trajectory to solve the problem was introduced. Further extension to the proposed GA was to use fuzzy motion control to apply it in dynamic environments as presented in [54]. All simulations in [52–54] were conducted on 20*20 maps with different obstacles ratio, and the start and the goal positions were fixed to the corners of the map, i.e., (0, 0) and (19, 19), respectively.

In [55], Cabreira et al. presented a GA-based approach for multi-agents in dynamic environments. Variable length chromosomes using binary encoding were adopted to represent the solutions where each gene is composed of three bits to represent a movement direction, i.e., up, left. Comparing the performance of the proposed GA with A* algorithm, the results show that the A* algorithm highly outperforms the GA, but the GA shows some improvements in the performance when the environment complexity grows. An extension for this work was introduced in [56], where the authors extended their simulations in complex and larger environments (up to 50*25) and propose new operator to improve the GA. In [57], an improved dual-population GA (IDPGA) algorithm for unmanned air vehicle (UAV) path planning was proposed. The feasible initial paths were generated based on multiple constraints. An additional population was used to preserve the diversity of GA, and they assigned varied functions to evaluate the individuals in that population which gives high values to the individuals with high distance from the main population. The main population will evolve as traditional GA to find good solutions. A comparative study using simulations of the proposed approach and conventional GA with single and multi (two)-populations was conducted. The authors proved that their approach was capable to enhance the GA abilities for both global and local searches. The IDPGA generates shorter paths and converges faster than the traditional GA with single population and multi-population.

2.4.3 Neural Networks

2.4.3.1 Overview

Artificial neural networks (ANNs or simply NNs) primarily inspired by biological neural networks [58, 59] are composed of a set of simple computational units (neurons), which are highly interconnected. Each neuron receives as input a linear combination of the outputs of neurons it is connected to. Then, a transition function (generally a sigmoid) is applied to this input in order to obtain the neuron's output.

Neural networks are mainly used to solve regression or classification problems, after an off-line supervised training phase on a representative learning labeled dataset. The simplest type of neural networks used for this aim is the **feed-forward NN**, also called **multilayer perceptron** (MLP), where neurons are organized in layers (an input layer, an output layer, and a set of hidden layers between them), with each layer fully connected to the next one. Other types of NN, such as **self-organizing map** (SOM) [60], can be used for clustering, after an unsupervised training phase on an unlabeled dataset.

These types of NN are not adapted to path planning problems, since a representative learning database cannot be provided for path planning. However, they have been used in other robot navigation applications; feed-forward neural networks have been used for environmental map building with high accuracy from infrared or sonar sensors using feed-forward NN [61, 62], global localization of mobile robots based on monocular vision [63], detecting and avoiding obstacles using a modular neural network [64], etc., while self-organizing maps have been used for recognizing landmarks in the process of initialization and re-calibration of the robot position [65] collision avoidance for an underwater vehicle [66], etc. A thorough survey of NN applications in robot navigation can be found in [67].

As for path planning of a point robot, different NN techniques with no learning process needed to be used. [68] successfully introduced the use of **Hopfield NN**, which is a form of recurrent neural networks [69]; any two neurons can be connected, without consideration for layers. The algorithm presented by [68] assumes that information about the position and shape of obstacles in the environment is not known beforehand, but appears during the path planning, for example, through real-time sensing. In addition, it guarantees an optimal solution (in the static case), if one exists. The optimal robot trajectory is generated without explicitly searching over the free workspace, without explicitly optimizing any global cost functions, without any prior knowledge of the dynamic environment, and without any learning procedures.

2.4.3.2 Basic Neural Network Algorithm

This topologically organized Hopfield-type neural network presented by [68] consists of a large collection of identical processing units (neurons), arranged in a d-dimensional cubic lattice, where d is the number of degrees of freedom of the robot. This lattice coincides with the **grid representation** of the state space, such that each state (including obstacles) is represented by a neuron.

The neurons are connected only to their z nearest neighbors (where $z = 4$ or 8, for a two-dimensional workspace) by excitatory and symmetric connections T_{ij}. The initial state q_{init} and the target state q_{targ} are represented by one node each. The obstacles are represented by neurons which activity is clamped to 0, while q_{targ} is clamped to 1. All other neurons have variable activity σ_i between 0 and 1 that changes due to inputs from other neurons in the network and due to external sensory input. The total input u_i to the neuron i is a weighted sum of activities from other neurons and of an external sensory input I_i:

$$u_i = \sum_j^N T_{ij}\sigma_j(t) + I_i \tag{2.4}$$

where T_{ij} is the connection between neuron i and neuron j, and N is the total number of neurons. These connections are excitatory ($T_{ij} > 0$), symmetric ($T_{ij} = T_{ji}$), and short range:

$$T_{ij} = \begin{Bmatrix} 0 \text{ if } \rho(i, j) < r \\ 1 \quad \text{otherwise} \end{Bmatrix} \tag{2.5}$$

where $\rho(i, j)$ is the Euclidean distance between neuron i and j.

In the case of time-discrete evolution, the activity of a neuron i (excluding the target and obstacles) is updated according to the following equation:

$$\sigma_t(t+1) = g\left(\sum_j^N T_{ij}\sigma_j(t) + I_i\right) \tag{2.6}$$

where g is a transition function that can be a sigmoid, or simply a linear function:

$$g(x) = a.x, \quad \text{with } a < \frac{1}{\beta} \tag{2.7}$$

β being the maximum number of neighbours. It is rigorously proven in [68] that, with the above conditions, the network dynamics converges to an equilibrium. The obtained path leads from one node of the lattice to the neighboring node with the largest activity and ends in the node with activity 1 (q_{target}). The number of steps and the length of the path obtained are proven to be minimal. Algorithm 4 presents a simple NN pseudocode.

2.4.3.3 Illustrative Example

Figure 2.5 shows a simple illustrative example of the NN basic algorithm. As can be seen in this example, in a static environment, an optimal path can be found even before the network settles into an equilibrium state. In fact, the optimal path is found when one of the neighboring neurons of the current neuron becomes activated (as is the case in step 7).

2.4.3.4 Other Variants of the Neural Network Algorithm

Yang and Meng [70] proposed a biologically inspired neural network approach, which can be seen as a generalization and an improvement of [68]'s Hopfield-type neural network. It is originally derived from Hodgkin and Huxley's biological membrane model [71], and it performs better in fast changing environments. The neural activity is a continuous analog signal and has both upper and lower bounds.

2.4.3.5 Neural Networks for Robot Path Planning: Literature Review

In [72], the author proposed a solution based on neural network to solve robot path planning for a static environment modeled as a grid. For each cell in the grid is associated a mesh processor that has five inputs (left, right, bottom bottom, an output and a bias term). To test the algorithm, the authors used 32*32 grid map. The algorithm converges after 168 iterations. Singh et al. [73] designed a neural network algorithm that enables a robot to move safely in an unknown environment with both static and dynamic obstacles. The neural network used is a four-layer perceptron. The inputs of the proposed neural network consist of left, right, and front obstacle distance (LOD, FOD, ROD, perspectively); the angle between the robot and the target (TA) and the output are robot steering angle. The information is received from an array of sensors. Compared with a fuzzy controller presented in [74], the algorithm dramatically reduces the computational time. Chen in [75] designed a robot navigation

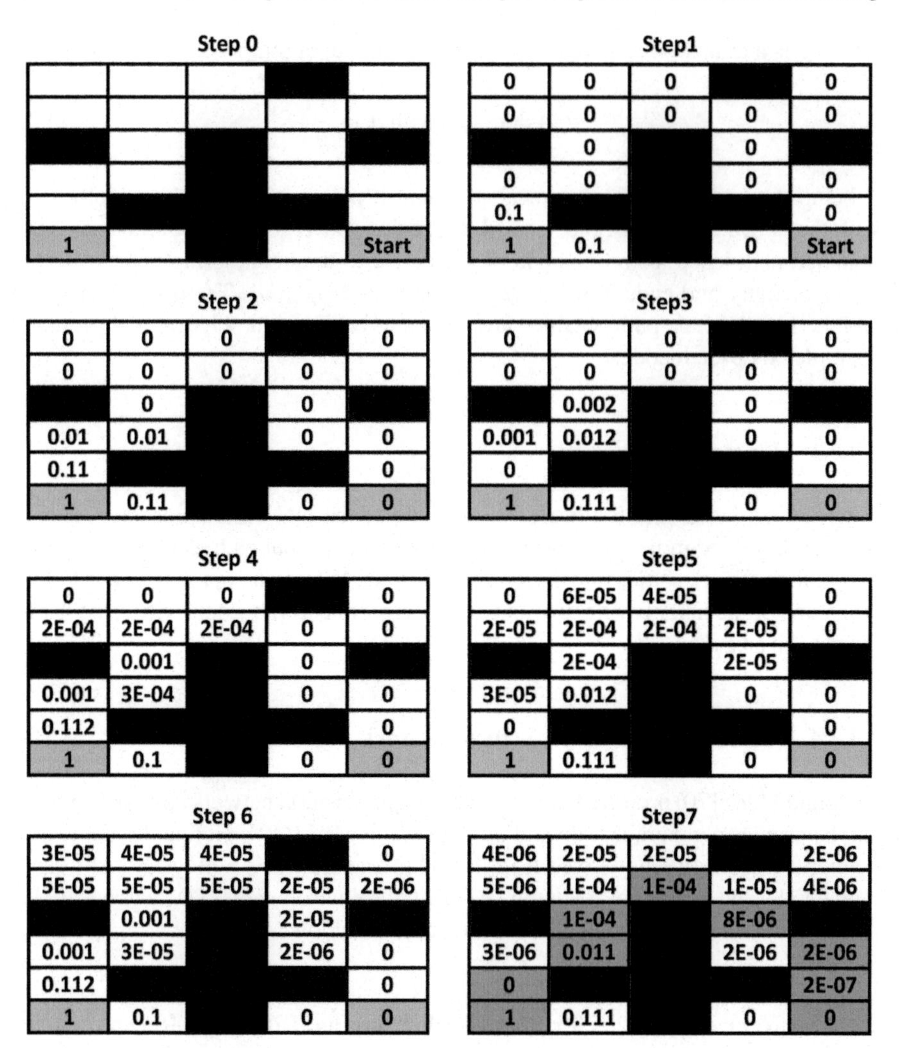

Fig. 2.5 Simple illustrative example of the NN basic algorithm in a static environment. Black cells represent obstacles. The maximum number of neighbours in this example is eight. And the transition function used is $g(x) = x/10$. The shortest path is obtained, in step 7, by following the neighboring node with the largest activity, at each move

system: A grid map building algorithm is proposed, and a modified pulse-coupled neural network (MPCNN) algorithm is proposed to find the optimal path for the robot. An experiment evaluation using an Arduino-based platform was conducted to demonstrate the effectiveness of the navigation system. The environment is a 9*9 grid map, the obstacles are static, and the robot stops when it receives IR waves indicating the presence of dynamic obstacles. The major drawbacks of this work are that it deals with small environments containing simple obstacle shapes and it

Algorithm 4. The Neural Network Algorithm

```
   // Initialisation:
 1 for i=1 to Nb_states do
 2 |   if i=Target then
 3 |   |   neuron_activity(i)=1;
 4 |   else
 5 |   |   neuron_activity(i)=0;
 6 |   end
 7 end
   // Iterative Update:
 8 while neuron_activity(Start)=0 do
 9 |   for i in {All States} - {Obstacles} - {Target} do
10 |   |   for j in Neighbourhood(i) do
11 |   |   |   neuron_activity(i)=neuron_activity(i)+neuron_activity(j);
12 |   |   end
13 |   |   neuron_activity(i)=transition_function(neuron_activity(i));
14 |   end
15 end
   // Path construction:
16 i=Start;
17 Path=[Start];
18 while i ≠ Target do
19 |   max_activity=neuron_activity(i);
   |   // Find the neuron with maximum activity in the
   |      neighbourhood of the current neuron:
20 |   for j in Neighbourhood(i) do
21 |   |   if neuron_activity(i) > max_activity then
22 |   |   |   neuron_activity(i)=1;
23 |   |   |   max_activity=neuron_activity(j);
24 |   |   |   next_state=j;
25 |   |   end
26 |   end
27 |   Add it to the path:;
28 |   Path=[Path ; next_state];
29 end
```

considers only one scenario (start/goal) in the experiment. Fuzzy modeling of the real robot's environment and a Hopfield neural network has been presented in [76], and the authors tested the algorithm in 400*300 grid map (only one scenario). They proved that Hopfield neural network fails to find the optimal solution and always get trapped in a suboptimal solution. To solve this issue, a combination of "Head-to-Tail" and "Tail-to-Head" procedures was designed in order to improve the final result of the algorithm. It has been argued that the proposed approach generates the optimal solution even in the presence of U-shaped obstacles. In [77], the authors proposed a NN technique for real-time collision-free path planning of a robot manipulator in a dynamic environment. A dynamic neural activity landscape was used to represent the dynamic environment. The authors showed that their algorithm is computationally efficient with a linear complexity that grows with the neural network size. Also, Reference [78] proposed the shortest path cellular neural network (SP-CNN) technique, which was shown to be effective for both static and dynamic environments. Each neuron in the NN only has local connection with its neighbors, and it has a two-layer architecture composed of a competitive layer and a dynamic layer. Extensive simulation results demonstrated how well the proposed approach was able to recognize the change of target tracking system in real time. In [79], the paper proposed a modified pulse-coupled neural network (MPCNN) model for collision-free path planning of mobile robots in dynamic environments, with real-time constraints. The key point is that the approach does not require any prior knowledge of target or barrier mobility pattern. Like in other works, simulation was used to demonstrate the effectiveness of the proposal.

2.4.4 *Ant Colony Optimization*

2.4.4.1 Overview

Ant colony optimization is metaheuristic, introduced by Marco Dorigo in his Ph.D. thesis in 1992 [80]. This approach is a part of the swarm intelligence[1]; it is inspired from the real-life behavior of a colony of ants, when looking for food, to find approximate solutions to difficult optimization problems.

ACO has been successfully used to solve several optimization problems, such as the traveling salesman problem (TSP) [81], the vehicle routing problem, the quadratic assignment.

2.4.4.2 ACO's Main Idea

In the nature, some ant species are able to select the shortest path, among a set of alternative routes, from their nest to the food source. This is accomplished by using *pheromone* deposits. In fact, ants lay on the ground a chemical trail, called pheromone, in the process of moving; this trail attracts other ants, which tend to follow the path that has the highest concentration of pheromone. After a certain period of time, the concentration of pheromone becomes greater along the shortest path and evaporates in the bad paths; this behavior is called *stigmergy*. If an obstacle interrupts their paths, the ants choose their paths in a probabilistic way at first, and then, after a period of time, the concentration of pheromone becomes higher in the shortest path and all ants choose this path. This process is illustrated in Fig. 2.6.

2.4.4.3 The ACO Algorithm

In general, the ACO algorithm follows a specific algorithmic scheme presented in Algorithm 5. At first, an initialization phase takes place during which the initial pheromone value and many other parameters are set. After that, a main loop is

[1]The swarm intelligence (SI) is the collective behavior of decentralized, self-organized systems, natural or artificial.

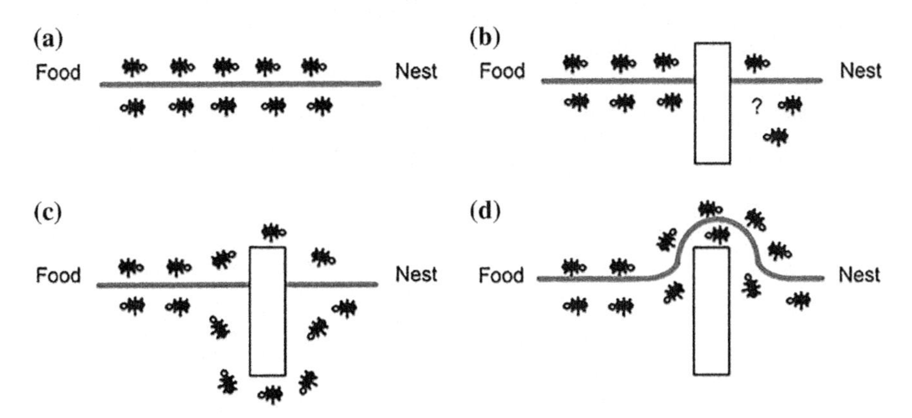

Fig. 2.6 **a** Ants in a pheromone trail between nest and food; **b** an obstacle interrupts the trail; **c** ants find two paths and go around the obstacles; **d** a new pheromone trail is formed along the shortest path

Algorithm 5. The Ant Colony Optimization Algorithm

```
1 Set parameters, initialize pheromone ;
2 while (termination condition not met) do
3 |   ConstructAntSolutions;
4 |   ApplyLocalSearch ;
5 |   %optional UpdatePheromones;
6 end
```

repeated until a termination condition is reached. In the beginning of the main loop, the ants build feasible solutions, which can be improved by applying a local search procedure. Finally, the quantity of pheromone is updated: Indeed, the pheromone quantity can either increase, as ants deposit pheromone on the components used during search, or decrease due to the pheromone evaporation from the unused components during search.

2.4.4.4 ACO Variants

In the literature, several ACO approaches have been proposed, which differ in some choices characterizing the ConstructAntSolutions and UpdatePheromones procedures. Table 2.1 is taken from [82], and it introduces a variety of ACO approaches:

Table 2.1 Different ACO Approaches

Approach	Authors	Year	Description
Ant system(AS)	Dorigo et al.	1991	Introduced above
Elitist AS	Dorigo et. al.	1992	Only the best ant deposits pheromone in each iteration
Ant-Q	Gambarella and Dorigo	1995	It is an extension of AS reinforced by Q-learning
Ant colony system	Dorigo and Gambarella	1996	The introduction of a pseudorandom proportional rule. The global updating rule is applied only to edges which belong to the best ant tour. A local pheromone updating rule is applied at the end of each iteration
MAX-MIN AS	Stutzle and Hoos	1996	The introduction of maximum and minimum pheromone amounts. The pheromone is deposited only on the best tour
Rank-based AS	Bullnheimer et. al.	1997	All solutions are ranked according to their costs. The amount of pheromone deposited is weighted for each solution with shorter paths deposit more pheromone than the solutions with longer paths
BWAS	Cordon et al.	2000	Three main variations as compared to AS: (1) the use of an aggressive update rule in which only the best ant deposits pheromone, (2) frequent re-inialization of the pheromone trails, (3) The introduction of the pheromone mutation concept
Hyper-cube AS	Blum et al.	2001	The presentation of a new pheromone update rule different from that of AS algorithm

2.4.4.5 Ant Colony Optimization for Robot Path Planning: Literature Review

Since its appearance, ACO has been extensively used to solve robot path planning. The use of ACO is relatively new as compared to other approaches. In what follows, we present some relevant works to give some insights on ACO for solving the path planning problem.

One of the earliest works is [83], as claimed in that paper, to present a solution to the path planning problem using an intensified ACO algorithm. The authors used a novel function in the beginning of the algorithm, instead of the transition rule probability (refer to Chap. 3), to calculate the probabilities of transition of the ants. Moreover, they have used a variable value of α and β that depend on time.

Very recently, [84] proposed a new ACO algorithm to solve the global path planning problem, which they called Heterogeneous ACO (HACO). The key differences with the traditional ACO algorithm are (1) the use of a new transition probability function, (2) a new rule of pheromone update rule, and (3) the proposal of the new concept of heterogeneous ants, meaning that ants have different characteristics exploited to optimize the search process of the solution. Simulation was conducted and showed the effectiveness of the heterogeneous ACO as compared to the traditional ACO in providing more accurate solutions.

Porta-Garcia et al. in [85] have presented a new solution for the path planning problem based on ACO approach, which they called "SACOdm" where d stands for distance and m for memory. One of the main contributions of this work is the inclusion of memory capabilities to the ants in order to avoid stagnation. Another contribution is the use of a fuzzy inference system (FIS) to evaluate the best paths found by the ants. The FIS considers two inputs, the distance of the generated path and the effort, i.e., the energy spent by the robot to make turns across the path, and one output, which is the cost of the path. The cost of the path is added to the length of the path, giving the final weight of the path. The SACOdm method was compared with the classical ACO method, and it has been shown that the former reduces the execution time up to 91%.

In paper [86], the authors proposed a hybrid algorithm that combines global and local path planning in dynamic environment that contains U- and V-shaped obstacles. In a first step, the algorithm generates an optimal path (off-line) toward the goal based on a modified ACO algorithm considering only the static obstacles. In the second step, the rolling windows technique is applied to avoid dynamic obstacles that appear while the robot navigates, following the path found in the first step, toward the goal position. The authors evaluated the performance of the modified ACO in terms of path length and convergence speed, and they compared them against the conventional ACO. They proved that the convergence speed of MACO is faster and the path length generated by MACO is shorter than CACO. Moreover, the different techniques of the rolling window enable the robot to avoid safely dynamic obstacles and also to re-plan its trajectory and choose the shortest one toward the goal position.

In [87], Zhang et al. proposed an improved ACO algorithm. The key differences with the traditional ACO algorithm are (1) the definition of an objective attraction function in the transition rule probability which is based on the attraction of the goal position. (2) The use of a modified pheromone update rule. The interesting point of this paper is the use of the RoboCupRescue simulation system to evaluate the efficiency of the algorithm. The authors compared the new algorithm with the conventional one in terms of path length and number of cycles, and they proved that their algorithm provides more accurate results. In [88], the authors presented a modified ACO algorithm to solve the global path planning problem in an indoor environment for a UAV. The workspace of the UAV is modeled as three-dimensional grid map. To overcome the shortcomings of the conventional ACO, the authors proposed a modified pheromone update rule: They proposed to update the quantity of pheromone only on the best path generated after an iteration of the algorithm. They also limited the quantity of pheromone to solve the problem of premature convergence of ACO. To evaluate the robot path, they added a climbing parameter in addition to the path cost, and they claimed that this parameter will help the robot to choose the best direction and bypass obstacles. It was demonstrated through simulation that the algorithm could produce good quality of solution in 10*10 grid map. In [89], Ganganath et al. proposed an off-line path planner for a mobile robot based on ant colony optimization algorithm with Gaussian functions (ACO 2 Gauss). The workspace of the mobile robot is a 3D hilly landscape. Unlike ordinary ACO algorithms, the proposed path planner provides the ants an extra flexibility in making routing decisions: An ant is allowed to select any point on the circumference of a circle with a radius of R as its next position. Each ant will select a set of circles to move toward the goal position. The pheromone is distributed on the center of the different circles as a 2D Gaussian function. In their simulation study, the authors compared the proposed algorithm against a preceding version called ACOGauss. It was demonstrated that the quality of solution of ACO-2-Gauss was improved as compared to ACO-Gauss path planner. Beside these old works, the path planning problem is still being investigated in recent works such as paper [79, 90]. In [90], the authors implemented the ant system algorithm for global robot path planning. Ten 20*20 grid maps that differ by the number and position of obstacles (from 1 to 10 obstacles) are used for simulation. Three handicaps of this research work: First, the use of small maps for experiments. Moreover, no comparison beeween the proposed technique and some other approaches has been conducted in this paper, and the major drawback of this work is the execution time of the algorithm which reaches, for example, 2017 seconds for 20*20 grid map containing only one circle obstacle. This hinders the practicability of this approach in real-world robot application which requires real-time execution. In [79], the authors proposed an improved ACO algorithm for the global path planning problem. To improve performance of the classical ACO, a modified transition rule probability was applied instead of the conventional one. It uses an attraction function to the goal position. They also used the pheromone updating rule based on the assignment rule of the wolf colony in order to avoid algorithm getting into local optimum and improve the convergence speed. They tested the algorithm in two different grid maps (20*20). The simulation results prove the efficiency of the proposed approach.

2.4.5 Hybrid Approaches

In addition to improving the different approaches proposed to solve the robot path planning problem by changing their basic characteristics, some works proposed improving the performance by combining two or more approaches together.

Some works proposed solutions based on the combination between ACO and GA. For instance, in [91] the authors presented a path planning method based on a hierarchical hybrid algorithm that aims at finding a path in a road traffic network. The network is divided into several subnetworks; ACO is applied on each subnetwork, and the best paths generated by ant colony optimization algorithms will be the initial population of genetic algorithms. In their simulation work, the authors proved that the novel algorithm can quickly and accurately find the optimal path in less number of iterations as compared to [92, 93]. Moreover, a combination between GA and ACO algorithms to solve the robot navigation problem was presented in [94]. A special function was proposed to improve the quality of the paths generated by the ants at the beginning of the algorithm. Crossover and mutation operators are applied to improve the quality of solution generated by the ACO algorithm and avoid falling into a local optimum. It was proved that the improved algorithm generates a better solution in terms of length, direction, and execution time or number of iterations as compared to the pure ACO algorithm. Geetha et al. [95] proposed a hybrid algorithm (ACOG) based on ACO and GA techniques. Darwinian reproduction, structure-preserving crossover, and structure-preserving mutation are the three genetic operations adopted to improve the efficiency of the ACO algorithm and to avoid falling into a local optimum. The authors claimed that their method is a multi-objective algorithm as it takes into consideration three different parameters: length, smoothness, and security of the path. In the simulation work, they compared the algorithm to literature [96], and they argued that the algorithm is able to generate near-optimal path.

In [97], the authors presented an ACO-GA-based algorithm. The novel algorithm introduces a modified transition rule probability function in the ACO algorithm that eliminates the parameter "distance between two positions" and also uses an additional parameters γ to control the behavior of the ants. The values of (α, β, γ) in the new transition rule probability are not constant as in the conventional one, and they evolve using the GA approach in order to obtain the best values to improve the accuracy of the algorithm to get the shortest path. The authors tested the new algorithm in three real environments, with different complexity, using the mobile robot (Khepera II). They compared their results with respect to the shortest path length obtained with Dijkstra's algorithm, and they succeed to obtain the shortest paths. Miao et al. [98] proposed a solution that combines the potential field method (PFM) and GA. First, PFM is used to create the initial population of GA which consists only of feasible paths, i.e., collision-free paths, and then, it is applied after each iteration of GA to all individuals in the generation. Different strategies were implemented in this work: the diversity strategy and the memory strategy. Under diversity strategy, they implemented the immigration, mutation, and crossover. The memory strategy is used to move the best solutions from current generation to the next generation. Using simulations, the algorithm was tested in 100*100 environments with different complexities (different numbers and sizes of obstacles with eight obstacles at maximum). The results show that the proposed algorithm has the ability to find the optimal solutions in the high diversity cases (i.e., high probabilities of crossover, mutation, and immigration). Also, the mutation operator has no impact in their algorithm.

In [52], Oleiwi et al. presented a hybrid approach based on GA, a modified search A* algorithm, and fuzzy logic. As a result, multi-objective optimization of free-collision path is generated. First, the modified GA and a modified search A* algorithm are applied to find the optimal path. After that, the global optimal path is considered as input for fuzzy motion controller to regenerate a time-based path. When unknown obstacles appear in the robot path, the fuzzy controller will decrease the speed of the robot. The performances are very impressive in the environment with dynamic obstacles.

2.4.6 Comparative Study of Heuristic and Exact Approaches

To the best of our knowledge, the only previous comparative study of exact and heuristic approaches is [99]. In this paper, six different approaches for the global path planning problem were evaluated: breadth-first Search, depth-first search, A^*, Moore–Dijkstra, neural approach, and GA. Three parameters were evaluated: the distance traveled by the robot, the number of visited waypoints, and the computational time (with and without initialization). Simulation was conducted in four environments, and it was demonstrated that GA outperforms the other approaches in terms of distance and execution time. Although the presented work evaluates the performance of some exact and heuristic techniques for path planning, there is still a need to evaluate them in large-scale environments as the authors limited their study to small-size environments (up to 40*40 grid maps). In this paper, we improve on this by considering a vast array of maps of different natures (rooms, random, mazes, video games) with a much larger scale up to 2000*2000 cells. In [55], Cabreira et al. presented a GA-based approach for multi-agents in dynamic environments. Variable length chromosomes using binary encoding were adopted to represent the solutions where each gene is composed of three bits to represent a motion direction, i.e., up, left, down. GA performance was compared to that of A* algorithm using simulations with NetLogo platform [100]. The results show that the A* algorithm highly outperforms GA, but GA shows some improvements in the performance when the environment complexity grows. An extension for this work was presented in [56], where the authors extended their simulations to complex and larger environments (up to 50*25) and proposed a new operator to improve the GA.

2.4.7 Comparative Study of Heuristic Approaches

In the literature, some research efforts have proposed and compared solutions for path planning based on heuristic approaches. For instance, in [37], a comparative study between trajectory-based metaheuristic and population-based metaheuristic has been conducted. TS, simulated annealing (SA), and GA were evaluated in a well-known static environment. The experiment was performed in the German University in Cairo (GUC) campus map. Four evaluation metrics were considered in the experimental study: the time required to reach the optimal path, the number of iterations required to reach the shortest path, the time per each iteration, and the best path length found after five iterations. It has been argued that SA performs better than the other planners in terms of running time, while TS was shown to provide the shorter path. In [101], the authors used two metaheuristics ACO and GA for solving the global path planning problem in static environments. The algorithms have been tested in three workspaces that have different complexities. Performances of both algorithms were evaluated and compared in terms of speed and number of iterations that each algorithm makes to find an optimal path. It was demonstrated that the ACO method was more effective than the GA method in terms of execution time and number of iterations. In [102], Grima et al. proposed two algorithms for path planning where the first is based on ACO and the second is based on GA. The authors compared both techniques on a real-world deployment of multiple robotic manipulators with specific spraying tools in an industrial environment. The authors claimed that both solutions provide very comparable results for small problem sizes, but when increasing the size and the complexity of the problem, the ACO-based algorithm produces a shorter path at the cost of a higher execution time, as compared to the GA-based algorithm. Four heuristic methods were compared in [103]: ACO, particle swarm optimization (PSO), GA, and a new technique called Quad harmony search (QHS) which is a combination between the Quad tree algorithm and the harmony search method. The Quad tree method is used to decompose the grid map in order to accelerate the search process, and the harmony method is used to find the optimal path. The authors demonstrated through simulation and experiments that QHS gives the best planning times in grid map with a lower percentage of obstacles. The work reported in [104] presents a comparison between two modified algebraic methods for path planning in static environments, the artificial potential field method enhanced with dummy loads and Voronoi diagram method enhanced with the Delaunay triangulation. The proposed algorithms were tested on 25 different cases (start/goal). For all test cases, the system was able to quickly determine the navigation path without falling into local minima. In the case of the artificial potential field method enhanced with dummy loads, the paths defined by the algorithm always avoided obstacles (the obstacles that cause local minima) by passing them in the most efficient way. The algorithm creates quite short navigation paths. Compared with Voronoi, this algorithm is computationally more expensive, but it finds optimal routes in 100% of the cases.

2.4.8 Comparative Study of Exact Methods

Other works compared solutions based on exact methods. The authors in [105] presented a comparative study of three A^* variants: D^*, Two Way D* (TWD^*), and E^*. The algorithms have been evaluated in terms of path characteristics (path cost, the number of points at which path changes directions, and the sum of all angles accumulated at the points of path direction changes), the execution time, and the number of iterations of the algorithm. They tested the planners on three different environments. The first is a grid map randomly generated of size 500*500, the second is a 165*540 free map, and the third is a 165*540 realistic map. The authors concluded that E^* and TWD^* produce shorter and more natural paths than D^*. However, D^* exhibits shorter runtime to find the best path. The interesting point of this paper is the test of the different algorithms on a real-world application using Pioneer 3DX mobile robot. The authors claimed that the three algorithms have a good real-time performance. In [106], the authors tackled the path planning problem in static and dynamic environments. They compared three different methods: a modified bug algorithm, the potential field method, and the A^* method. The authors concluded that the modified bug algorithm is an effective technique for robot path planning for both static and dynamic maps. A^* is the worst technique as it requires a large computational effort. Eraghi et al. in [107] compared A^*, Dijkstra, and HCTNav which have been proposed in [108], and HCTNav algorithm is designed for low-resource robots navigating in binary maps. HCTNav's concept is to combine the shortest path search principles of the deterministic approaches with the obstacle detection and avoidance techniques of the reactive approaches. They evaluated the performance of the three algorithms in terms of path length, execution time, and memory usage. The authors argued that the results in terms of path length of the three algorithms are very similar. However, HCTNav needs less computational resources, especially less memory. Thus, HCTNav can be a good alternative for navigation in low-cost robots. In [109], both global and local path planning problems are tackled. The authors compared five different methods: A^*, Focused D^*, Incremental Phi^*, Basic Theta*, and jump point search (JPS). They concluded that the JPS algorithm achieves near-optimal paths very quickly as compared to the other algorithms. Thus, if the real-time character is imperative in the robot application, JPS is the best choice. If there is no requirement of a real time and the length of path plays a big role, then Basic Theta* algorithm is recommended. Focused D* and Incremental Phi^* are not appropriate for static environments. They can be used in dynamic environments with a small amount of obstacles. In [110], the authors compared two path planning algorithms that have the same computational complexity $\mathcal{O}(nlog(n))$ (where n is the number of grid point): fast marching method (FMM) and A^*. They tested the two algorithms on grid maps of sizes 40*40 up to 150*150. The authors claimed that A^* is faster than the other planners and it generates continuous but not smooth path, while FMM generates the best path (smoothest and shortest) as the resolution of the map gets finer and finer. Other research works addressed the path planning problem in unknown environments. In [111], the authors made a comparison study of five path planning algorithms in unstructured environments (A^*, RRT, PRM, artificial potential field (APF), and the proposed free configuration eigen-space (FCE) path planning method). They analyzed the performance of the algorithms in terms of computation time needed to find a solution, the distance traveled, and the amount of turning by the autonomous mobile robot, and they showed that the PRM technique provides a shorter path than RRT, but RRT is faster and produces a smooth path with minimum direction changes. A^* generates an optimal path, but its computational time is high and the clearance space from the obstacle is low. The APF algorithm suffers from local minima problem. In case of FCE, the path length and turning value are comparatively larger than all other methods. The authors considered that in case of planning in unknown environments, a good path is relatively short, keeps some clearance distance from the obstacles, and is smooth. They concluded that APF and the proposed FCE techniques are better with respect to these attributes. The work reported in [112] evaluated the performance of A*, Dijkstra, and breadth-first search to find out the most suitable path planning algorithm for rescue operation. The three methods are compared for two cases: for one starting-one goal cells and for one starting-multi-goal cells in 256*256 grid in terms of path length, number of explored nodes, and

CPU time. A* was found to be the best choice in case of maps containing obstacles. However, for free maps, breadth-first search is the best algorithm for both cases (one starting-one goal cell and one starting-multi-goal cells) if the execution time is the selection criteria. A* can be a better alternative if the memory consumption is the selection criteria.

2.5 Conclusion

In this chapter, we reviewed some research works presented in the literature. These works pertain to two main categories: exact methods such as (A^*) and Dijkstra an metaheuristics including Tabu Search, genetic algorithms, ant colony optimization, etc. This diversity and difference raise the complex challenge of choosing the best algorithm for the path planning problem which is the main objective of the iroboapp research project. This makes the subject of the next chapter.

References

1. Latombe, Jean claude. 1991. *Robot motion planning*. The Springer International Series in Engineering and Computer Science.
2. Šeda, Miloš. 2007. Roadmap methods versus cell decomposition in robot motion planning. In *Proceedings of the 6th WSEAS international conference on signal processing, robotics and automation*, 127–132. World Scientific and Engineering Academy and Society (WSEAS).
3. Yan, Zhi, Nicolas Jouandeau, and Arab Ali Cherif. 2013. Acs-prm: Adaptive cross sampling based probabilistic roadmap for multi-robot motion planning. In *Intelligent autonomous systems 12*, 843–851. Springer.
4. Nazif, Ali Nasri, Alireza Davoodi, and Philippe Pasquier. 2010. Multi-agent area coverage using a single query roadmap: A swarm intelligence approach. In *Advances in practical multi-agent systems*, 95–112. Springer.
5. Rosell, Jan, and Pedro Iniguez. 2005. Path planning using harmonic functions and probabilistic cell decomposition. In *Proceedings of the 2005 IEEE international conference on robotics and automation. ICRA 2005*, 1803–1808. IEEE.
6. Cosío, F., M.A.Padilla Arambula, and Castañeda. 2004. Autonomous robot navigation using adaptive potential fields. *Mathematical and Computer Modelling* 40 (9–10): 1141–1156.
7. Sfeir, Joe, Maarouf Saad, and Hamadou Saliah-Hassane. 2011. An improved artificial potential field approach to real-time mobile robot path planning in an unknown environment. In *2011 IEEE international symposium on robotic and sensors environments (ROSE)*, 208–213. IEEE.
8. Kim, Dong Hun. 2009. Escaping route method for a trap situation in local path planning. *International Journal of Control, Automation and Systems* 7 (3): 495–500.
9. Hart, Peter E., Nils J. Nilsson, and Bertram Raphael. 1968. A formal basis for the heuristic determination of minimum cost paths. *IEEE Transactions on Systems Science and Cybernetics* 4 (2): 100–107.
10. Choubey, Neha, and Mr. Bhupesh Kr. Gupta. 2013. Analysis of working of dijkstra and a* to obtain optimal path. *International Journal of Computer Science and Management Research* 2: 1898–1904. <pagination /><pagination />

11. Potamias, Michalis, Francesco Bonchi, Carlos Castillo, and Aristides Gionis. 2009. Fast shortest path distance estimation in large networks. In *Proceedings of the 18th ACM conference on Information and knowledge management*, 867–876. ACM.
12. Jigang, Wu, and Pingliang Han, George Rosario Jagadeesh, and Thambipillai Srikanthan. 2010. Practical algorithm for shortest path on large networks with time-dependent edge-length. In 2010 2nd international conference on computer engineering and technology (ICCET), vol. 2, 57–60. China: Chengdu.
13. Kanoulas, Evangelos, Yang Du, Tian Xia, and Donghui Zhang. 2006. Finding fastest paths on a road network with speed patterns. In *Proceedings of the 22nd International Conference on Data Engineering, ICDE'06*, 10. IEEE.
14. Dijkstra, Edsger W. 1959. A note on two problems in connexion with graphs. *Numerische mathematik* 1 (1): 269–271.
15. Pearl Judea. 1984. *Heuristics: Intelligent search strategies for computer problem solving*.
16. Pohl, Ira. 1970. First results on the effect of error in heuristic search. *Machine Intelligence* 5: 219–236.
17. Pohl, Ira. 1973. The avoidance of (relative) catastrophe, heuristic competence, genuine dynamic weighting and computational issues in heuristic problem solving. In *Proceedings of the 3rd international joint conference on artificial intelligence*, 12–17. Morgan Kaufmann Publishers Inc.
18. Köll, Andreas, and Hermann Kaindl. 1992. A new approach to dynamic weighting. In *Proceedings of the tenth European conference on artificial intelligence (ECAI-92)*, 16–17, Vienna, Austria.
19. Harabor, Daniel Damir, Alban Grastien, et al. 2011. Online graph pruning for pathfinding on grid maps. In *AAAI*.
20. Cazenave, Tristan. 2006. Optimizations of data structures, heuristics and algorithms for pathfinding on maps. In *2006 IEEE symposium on computational intelligence and games*, 27–33. IEEE.
21. Korf, Richard E. 1985. Depth-first iterative-deepening: An optimal admissible tree search. *Artificial Intelligence* 27 (1): 97–109.
22. Antsfeld, Leonid, Daniel Damir Harabor, Philip Kilby, and Toby Walsh. 2012. Transit routing on video game maps. In *AIIDE*, 2–7.
23. Botea, Adi, et al. Ultra-fast optimal path finding without runtime search. In *AIIDE*.
24. Likhachev, Maxim, Geoffrey J Gordon, and Sebastian Thrun. 2004. Ara*: Anytime a* with provable bounds on sub-optimality. In *Advances in Neural Information Processing Systems*, 767–774.
25. Likhachev, Maxim, David I Ferguson, Geoffrey J Gordon, Anthony Stentz, and Sebastian Thrun. 2005. Anytime dynamic a*: An anytime, replanning algorithm. In *ICAPS*, 262–271.
26. Koenig, Sven, and Maxim Likhachev. 2002. D* lite. In *Proceedings of the eighteenth national conference on artificial intelligence (AAAI)*, 476–483.
27. Berg, Jur Van Den, Rajat Shah, Arthur Huang, and Ken Goldberg. 2011. Ana*: Anytime nonparametric a*. In *Proceedings of twenty-fifth AAAI conference on artificial intelligence (AAAI-11)*.
28. Glover, Fred. 1986. Future paths for integer programming and links to artificial intelligence. *Computers and Operations Research*, 13 (5): 533 – 549. Applications of Integer Programming.
29. Osman, Ibrahim H., and Gilbert Laporte. 1996. Metaheuristics: A bibliography. *Annals of Operations Research* 63 (5): 511–623.
30. Voss, Stefan, Ibrahim H. Osman, and Catherine Roucairol (eds.). 1999. *Meta-Heuristics: Advances and trends in local search paradigms for optimization*. Norwell, MA, USA: Kluwer Academic Publishers.
31. Mohanty, Prases K., and Dayal R. Parhi. 2013. Controlling the motion of an autonomous mobile robot using various techniques: A review. *Journal of Advance of Mechanical Engineering* 1 (1): 24–39.
32. Kirkpatrick, S., C.D. Gelatt, and M.P. Vecchi. 1983. Optimization by simulated annealing. *Science* 220 (4598): 671–680.

33. Glover, Fred. 1989. Tabu search-part i. *ORSA Journal on Computing* 1 (3): 90–206.
34. Glover, Fred. 1990. Tabu search-part ii. *ORSA Journal on Computing* 2 (1): 4–32.
35. Masehian, Ellips, and MR Amin-Naseri. 2006. A tabu search-based approach for online motion planning. In *IEEE International Conference on Industrial Technology. ICIT 2006*, 2756–2761. IEEE.
36. Masehian, Ellips, and Mohammad Reza Amin-Naseri. 2008. Sensor-based robot motion planning-a tabu search approach. *IEEE Robotics and Automation Magazine* 15 (2):
37. Hussein, Ahmed, Heba Mostafa, Mohamed Badrel-din, Osama Sultan, and Alaa Khamis. 2012. Metaheuristic optimization approach to mobile robot path planning. In *2012 International Conference on Engineering and Technology (ICET)*, 1–6. IEEE.
38. Khaksar, Weria, Tang Sai Hong, Mansoor Khaksar, and Omid Reza Esmaeili Motlagh. 2012. Sampling-based tabu search approach for online path planning. *Advanced Robotics* 26 (8–9): 1013–1034.
39. Wei, Hongxing, Bin Wang, Yi Wang, Zili Shao, and Keith C.C. Chan. 2012. Staying-alive path planning with energy optimization for mobile robots. *Expert Systems with Applications* 39 (3): 3559–3571.
40. Wang, Tianmiao, Bin Wang, Hongxing Wei, Yunan Cao, Meng Wang, and Zili Shao. 2008. Staying-alive and energy-efficient path planning for mobile robots. In *American control conference*, 868–873.
41. Tang, Kit-Sang, Kim-Fung Man, Sam Kwong, and Qun He. 1996. Genetic algorithms and their applications. *IEEE Signal Processing Magazine* 13 (6): 22–37.
42. Yongnian, Zhou, Zheng Lifang, and Li Yongping. 2012. An improved genetic algorithm for mobile robotic path planning. In *2012 24th Chinese control and decision conference (CCDC)*, 3255–3260. IEEE.
43. Jianguo, Wang, Ding Biao, Miao Guijuan, Bao Jianwu, and Yang Xuedong. 2012. Path planning of mobile robot based on improving genetic algorithm. In *Proceedings of the 2011 international conference on informatics, cybernetics, and computer engineering (ICCE2011) November 1920, 2011, Melbourne, Australia*, vol. 112, ed. Liangzhong Jiang, 535–542. Advances in Intelligent and Soft Computing. Berlin: Springer.
44. Zhao, Jie, Lei Zhu, Gangfeng Liu, Gang Liu, and Zhenfeng Han. 2009. A modified genetic algorithm for global path planning of searching robot in mine disasters. In *ICMA 2009 international conference on mechatronics and automation*, 4936–4940.
45. Nearchou, Andreas C. 1998. Path planning of a mobile robot using genetic heuristics. *Robotica* 16: 575–588.
46. Lee, J., B.-Y. Kang, and D.-W. Kim. 2013. Fast genetic algorithm for robot path planning. *Electronics Letters* 49 (23): 1449–1451.
47. Sedighi, Kamran H., Theodore W. Manikas, Kaveh Ashenayi, and Roger L. Wainwright. 2009. A genetic algorithm for autonomous navigation using variable-monotone paths. *International Journal of Robotics and Automation* 24 (4): 367.
48. Karami, Amir Hossein, and Maryam Hasanzadeh. 2015. An adaptive genetic algorithm for robot motion planning in 2d complex environments. *Computers and Electrical Engineering* 43: 317–329.
49. Liu, Shuhua, Yantao Tian, and Jinfang Liu. 2004. Multi mobile robot path planning based on genetic algorithm. In *WCICA 2004 fifth world congress on intelligent control and automation*, vol. 5, 4706–4709.
50. Rastogi, Shivanshu, and Vikas Kumar. 2011. An approach based on genetic algorithms to solve the path planning problem of mobile robot in static environment. *MIT International Journal of computer science and information technology* 1: 32–35.
51. Tamilselvi, D., S. Mercy Shalinie, A. Fathima Thasneem, and S. Gomathi Sundari. 2012. Optimal path selection for mobile robot navigation using genetic algorithm in an indoor environment. In *Advanced Computing, Networking and Security*, vol. 7135, ed. P. Santhi Thilagam, Pais AlwynRoshan, K. Chandrasekaran, and N. Balakrishnan, 263–269. Lecture Notes In Computer Science. Berlin: Springer.

52. Oleiwi, Bashra K., Hubert Roth, and Bahaa I. Kazem. 2014. Modified genetic algorithm based on a* algorithm of multi objective optimization for path planning. *Jounal of Automation and Control Engineering* 2 (4): 357–362.
53. Oleiwi, Bashra Kadhim, Hubert Roth, and Bahaa I. Kazem. 2014. Multi objective optimization of path and trajectory planning for non-holonomic mobile robot using enhanced genetic algorithm. In *Neural networks and artificial intelligence*, vol. 440, ed. Vladimir Golovko, and Akira Imada, 50–62. Communications in Computer and Information Science: Springer International Publishing.
54. Oleiwi, Bashra Kadhim, Rami Al-Jarrah, Hubert Roth, and Bahaa I. Kazem. 2014. Multi objective optimization of trajectory planning of non-holonomic mobile robot in dynamic environment using enhanced ga by fuzzy motion control and a*. In *Neural Networks and Artificial Intelligence*, eds. Vladimir Golovko and Akira Imada, vol. 440, 34–49. Communications in Computer and Information Science. Springer International Publishing.
55. Cabreira, T.M., G.P. Dimuro, and M.S. de Aguiar. 2012. An evolutionary learning approach for robot path planning with fuzzy obstacle detection and avoidance in a multi-agent environment. In *2012 third Brazilian workshop on social simulation (BWSS)*, 60–67.
56. Cabreira, T.M., M.S. de Aguiar, and G.P. Dimuro. 2013. An extended evolutionary learning approach for multiple robot path planning in a multi-agent environment. In *2013 IEEE congress on evolutionary computation (CEC)*, 3363–3370.
57. Xiao-Ting, Ji, Xie Hai-Bin, Zhou Li, and Jia Sheng-De. 2013. Flight path planning based on an improved genetic algorithm. In *2013 third international conference on intelligent system design and engineering applications (ISDEA)*, 775–778.
58. Rosenblatt, Frank. 1958. The perceptron: A probabilistic model for information storage and organization in the brain. *Psychological Review* 65 (6): 386.
59. Haykin, Simon. 1998. *Neural networks: A comprehensive foundation*, 2nd ed. Upper Saddle River, NJ, USA: Prentice Hall PTR.
60. Kohonen, Teuvo (ed.). 2001. *Self-organizing maps*. Berlin: Springer.
61. Thrun, Sebastian B. 1993. Exploration and model building in mobile robot domains. In *IEEE international conference on neural networks*, 175–180. IEEE.
62. Kim, Heon-Hui, Yun-Su Ha, and Gang-Gyoo Jin. 2003. A study on the environmental map building for a mobile robot using infrared range-finder sensors. In *Proceedings of the 2003 IEEE/RSJ international conference on intelligent robots and systems, IROS 2003*, vol. 1, 711–716. IEEE.
63. Zou, Anmin, Zengguang Hou, Lejie Zhang, and Min Tan. 2005. A neural network-based camera calibration method for mobile robot localization problems. In *International symposium on neural networks*, 277–284. Springer.
64. Silva, Catarina, Manuel Crisostomo, and Bernardete Ribeiro. 2000. Monoda: a neural modular architecture for obstacle avoidance without knowledge of the environment. In *Proceedings of the IEEE-INNS-ENNS international joint conference on neural networks, IJCNN 2000*, vol. 6, 334–339. IEEE.
65. Hu, Huosheng, and Dongbing Gu. 1999. Landmark-based navigation of mobile robots in manufacturing. In *Proceedings of the 7th IEEE international conference on emerging technologies and factory automation, ETFA'99*, vol. 1, 121–128. IEEE.
66. Ishii, Kazuo, Syuhei Nishida, Keisuke Watanabe, and Tamaki Ura. 2002. A collision avoidance system based on self-organizing map and its application to an underwater vehicle. In *7th international conference on control, automation, robotics and vision, ICARCV 2002*, vol. 2, 602–607. IEEE.
67. Zou, An-Min, Zeng-Guang Hou, Fu Si-Yao, and Min Tan. 2006. Neural networks for mobile robot navigation: a survey. In *Advances in Neural Networks-ISNN*, 1218–1226.
68. Glasius, Roy, C.A.M. Andrzej Komoda, and Stan, and Gielen. 1995. Neural network dynamics for path planning and obstacle avoidance. *Neural Networks* 8 (1): 125–133.
69. Hopfield, John J. 1987. Neural networks and physical systems with emergent collective computational abilities. In *Spin glass theory and beyond: An introduction to the replica method and its applications*, 411–415. World Scientific.

70. Yang, Simon X. and Max Meng. 2001. Neural network approaches to dynamic collision-free trajectory generation. *IEEE Transactions on Systems, Man, and Cybernetics, Part B (Cybernetics)*, 31 (3): 302–318.
71. Hodgkin, A.L., and A.F. Huxley. 1952. A quantitative description of membrane current and its application to conduction and excitation in nerve. *The Journal of Physiology* 117 (4): 500–544.
72. Chan, H.T., K.S. Tam, and N.K. Leung. 1993. A neural network approach for solving the path planning problem. In *1993 IEEE international symposium on circuits and systems, ISCAS'93*, 2454–2457. IEEE.
73. Singh, Mukesh Kumar, and Dayal R. Parhi. 2009. Intelligent neuro-controller for navigation of mobile robot. In *Proceedings of the international conference on advances in computing, communication and control*, 123–128. ACM.
74. Pradhan, Saroj Kumar, Dayal Ramakrushna Parhi, and Anup Kumar Panda. 2009. Fuzzy logic techniques for navigation of several mobile robots. *Applied Soft Computing* 9 (1): 290–304.
75. Chen, Yi-Wen, and Wei-Yu Chiu. 2015. Optimal robot path planning system by using a neural network-based approach. In *2015 international automatic control conference (CACS)*, 85–90. IEEE.
76. Sadati, Nasser and Javid Taheri. 2002. Solving robot motion planning problem using hopfield neural network in a fuzzified environment. In *Proceedings of the 2002 IEEE international conference on fuzzy systems, FUZZ-IEEE'02*, vol. 2, 1144–1149. IEEE.
77. Simon, X., and Yang and Max Meng. 2000. An efficient neural network approach to dynamic robot motion planning. *Neural Networks* 13: 143–148.
78. Cao, Yan, Xiaolan Zhou, Shuai Li, Feng Zhang, Xinwei Wu, Aomei Li, and Lei Sun. 2010. Design of path planning based cellular neural network. In *2010 8th world congress on intelligent control and automation (WCICA)*, 6539–6544. IEEE.
79. Hong, Qu, Simon X. Yang, Allan R. Willms, and Zhang Yi. 2009. Real-time robot path planning based on a modified pulse-coupled neural network model. *IEEE Transactions on Neural Networks* 20 (11): 1724–1739.
80. Dorigo, Marco, Mauro Birattari, and Thomas Sttzle. 2006. Ant colony optimization-artificial ants as a computational intelligence technique. *IEEE Computational Intelligence Magazine* 1: 28–39.
81. Dorigo, Marco, and Luca Maria Gambardella. 1997. Ant colony system: A cooperative learning approach to the traveling salesman problem. *IEEE Transactions on evolutionary computation* 1 (1): 53–66.
82. Dorigo, Thomas Sttzle, and Marco. 2004. *Ant colony optimization*. Cambridge, Massachusetts London, England: The MIT Press.
83. Fan, Xiaoping, Xiong Luo, Sheng Yi, Shengyue Yang, and Heng Zhang. 2003. Optimal path planning for mobile robots based on intensified ant colony optimization algorithm. In *Proceedings of the 2003 IEEE international conference on robotics, intelligent systems and signal processing*, vol. 1, 131–136. IEEE.
84. Lee, Joon-Woo, Young-Im Choy, Masanori Sugisakaz, and Ju-Jang Lee. 2010. Study of novel heterogeneous ant colony optimization algorithm for global path planning. In *2010 IEEE international symposium on industrial electronics (ISIE)*, 1961–1966. IEEE.
85. Porta Garcia, M.A., Oscar Montiel, Oscar Castillo, Roberto Sepúlveda, and Patricia Melin. 2009. Path planning for autonomous mobile robot navigation with ant colony optimization and fuzzy cost function evaluation. *Applied Soft Computing* 9 (3): 1102–1110.
86. Dong-Shu, Wang, and Yu Hua-Fang. 2011. Path planning of mobile robot in dynamic environments. In *2011 2nd international conference on intelligent control and information processing (ICICIP)*, vol. 2, 691–696. IEEE.
87. Zhang, Xiaoyong, Min Wu, Jun Peng, and Fu Jiang. 2009. A rescue robot path planning based on ant colony optimization algorithm. In *International conference on information technology and computer science, ITCS 2009*, vol. 2, 180–183. IEEE.
88. He, Yufeng, Qinghua Zeng, Jianye Liu, Guili Xu, and Xiaoyi Deng. 2013. Path planning for indoor uav based on ant colony optimization. In *2013 25th Chinese control and decision conference (CCDC)*, 2919–2923. IEEE.

89. Ganganath, Nuwan, and Chi-Tsun Cheng. 2013. A 2-dimensional aco-based path planner for off-line robot path planning. In *2013 international conference on cyber-enabled distributed computing and knowledge discovery (CyberC)*, 302–307. IEEE.
90. Yee, Zi Cong, and S.G. Ponnambalam. 2009. Mobile robot path planning using ant colony optimization. In *IEEE/ASME international conference on advanced intelligent mechatronics, AIM 2009*, 851–856. IEEE.
91. Ma, Yong-jie, and Wen-jing Hou. 2010. Path planning method based on hierarchical hybrid algorithm. In *2010 international conference on computer, mechatronics, control and electronic engineering (CMCE)*, vol. 1, 74–77. IEEE.
92. Qing, L.I., Wei Zhang, Yi-xin Yin, and Zhi-liang Wang. 2006. An improved genetic algorithm for optimal path planning. *Journal of Information and Control*, 444–447.
93. Xu, Jing-Rong, Yun Li, Hai-Tao Liu, and Pan Liu. 2008. Hybrid genetic ant colony algorithm for traveling salesman problem. *Journal of Computer Applications*, 2084–2112.
94. Gao, Meijuan, Jin Xu, and Jingwen Tian. 2008. Mobile robot global path planning based on improved augment ant colony algorithm. In *Second international conference on genetic and evolutionary computing, WGEC'08*, 273–276. IEEE.
95. Geetha, S., G. Muthu Chitra, and V. Jayalakshmi. 2011. Multi objective mobile robot path planning based on hybrid algorithm. In *2011 3rd international conference on electronics computer technology (ICECT)*, vol. 6, 251–255. IEEE.
96. Zhou, Wang, Zhang Yi, and Yang Ruimin. 2008. Mobile robot path planning based on genetic algorithm. *Microcomputer Information* 24 (26): 187–189.
97. Garro, Beatriz A., Humberto Sossa, and Roberto A. Vazquez. 2007. Evolving ant colony system for optimizing path planning in mobile robots. In *Electronics, robotics and automotive mechanics conference, CERMA 2007*, 444–449. IEEE.
98. Miao, Yun-Qian, Alaa Khamis, Fakhreddine Karray, and Mohamed Kamel. 2011. A novel approach to path planning for autonomous mobile robots. *International Journal on Control and Intelligent Systems* 39 (4): 1–27.
99. Randria, Iadaloharivola, Mohamed Moncef Ben Khelifa, Moez Bouchouicha, and Patrick Abellard. 2007. A comparative study of six basic approaches for path planning towards an autonomous navigation. In *33rd annual conference of the IEEE industrial electronics society, IECON 2007*, 2730–2735. IEEE.
100. Tisue, Seth, and Uri Wilensky. 2004. Netlogo: A simple environment for modeling complexity. In *International conference on complex systems*, vol. 21, 16–21. Boston, MA.
101. Sariff, Nohaidda Binti, and Norlida Buniyamin. 2009. Comparative study of genetic algorithm and ant colony optimization algorithm performances for robot path planning in global static environments of different complexities. In *2009 IEEE international symposium on computational intelligence in robotics and automation (CIRA)*, 132–137. IEEE.
102. Tewolde, Girma S., and Weihua Sheng. 2008. Robot path integration in manufacturing processes: Genetic algorithm versus ant colony optimization. *IEEE Transactions on Systems, Man, and Cybernetics-Part A: Systems and Humans* 38 (2): 278–287.
103. Koceski, Saso, Stojanche Panov, and Natasa Koceska. 2014. Pierluigi Beomonte Zobel, and Francesco Durante. A novel quad harmony search algorithm for grid-based path finding. *International Journal of Advanced Robotic Systems* 11 (9): 144.
104. Gomez, Edwar Jacinto, Fernando Martinez Santa, and Fredy Hernan Martinez Sarmiento. 2013. A comparative study of geometric path planning methods for a mobile robot: potential field and voronoi diagrams. In *2013 II international congress of engineering mechatronics and automation (CIIMA)*, 1–6. IEEE.
105. Čikeš, Mijo, Marija akulović, and Ivan Petrović. 2011. The path planning algorithms for a mobile robot based on the occupancy grid map of the environment a comparative study. In *2011 XXIII international symposium on information, communication and automation technologies (ICAT)*, 1–8. IEEE.
106. Haro, Felipe, and Miguel Torres. 2006. A comparison of path planning algorithms for omnidirectional robots in dynamic environments. In *IEEE 3rd Latin American robotics symposium, LARS'06*, 18–25. IEEE.

107. Eraghi, Nafiseh Osati, Femando Lopez-Colino, Angel De Castro, and Javier Garrido. Path length comparison in grid maps of planning algorithms: Hctnav, a and dijkstra. In *2014 Conference on design of circuits and integrated circuits (DCIS)*, 1–6. IEEE.
108. Pala, Marco, Nafiseh Osati Eraghi, Fernando López-Colino, Alberto Sanchez, Angel de Castro, and Javier Garrido. 2013. Hctnav: A path planning algorithm for low-cost autonomous robot navigation in indoor environments. *ISPRS International Journal of Geo-Information* 2 (3): 729–748.
109. Duchoň, František, Peter Hubinskỳ, Andrej Babinec, Tomáš Fico, and Dominik Huňady. 2014. Real-time path planning for the robot in known environment. In *2014 23rd International Conference on robotics in Alpe-Adria-Danube region (RAAD)*, 1–8. IEEE.
110. Chiang, Chia Hsun, Po Jui Chiang, Jerry Chien-Chih Fei, and Jin Sin Liu. 2007. A comparative study of implementing fast marching method and a* search for mobile robot path planning in grid environment: Effect of map resolution. In *IEEE workshop on advanced robotics and its social impacts, ARSO 2007*, 1–6. IEEE.
111. Zaheer, Shyba, M. Jayaraju, and Tauseef Gulrez. 2015. Performance analysis of path planning techniques for autonomous mobile robots. In *2015 IEEE international conference on electrical, computer and communication technologies (ICECCT)*, 1–5. IEEE.
112. Al-Arif, S., A. Ferdous, and S. Nijami. 2012. Comparative study of different path planning algorithms: A water based rescue system. *International Journal of Computer Applications*, 39.

Chapter 3
Design and Evaluation of Intelligent Global Path Planning Algorithms

Abstract Global path planning is a crucial component for robot navigation in map-based environments. It consists in finding the shortest path between start and goal locations. The analysis of existing literature in Chap. 2 shows two main approaches commonly used to address the path planning problem: (1) exact methods and (2) heuristic methods. A* and Dijkstra are known to be the most widely used exact methods for mobile robot global path planning. On the other hand, several heuristic methods based on ant colony optimization (ACO), genetic algorithms (GA), Tabu Search (TS), and hybrid approaches of both have been proposed in the literature. One might wonder which of these methods is more effective for the robot path planning problem. Several questions also arise: Do exact methods consistently outperform heuristic methods? If so, why? Is it possible to devise more powerful hybrid approaches using the advantages of exact and heuristics methods? To the best of our knowledge, there is no comprehensive comparison between exact and heuristic methods in solving the path planning problem. This chapter fills the gap, addresses the aforementioned research questions, and proposes a comprehensive simulation study of exact and heuristic global path planners to identify the more appropriate technique for the global path planning.

3.1 Introduction

As previously discussed in Chap. 2, numerous research initiatives, aiming at providing different solutions for global path planning problem, have emerged. Different approaches to design these solutions have been attempted which can be widely classified into two main categories: (1) exact methods such as A_* and Dijkstra; (2) heuristic methods such as genetic algorithm (GA), Tabu Search (TS), and ant colony optimization (ACO). In the context of the iroboapp research project [1], we investigate, in this chapter, the capabilities of a variety of algorithms, both exact and heuristic as well as a relaxed version of A^*, used to solve the global path planning. We performed a comprehensive comparative study between the algorithms. In this work, our concern is not only optimality but also the real-time execution for large and complex grid map-based environments (up to 2000 * 2000 grid maps) as this is an important

A. Koubaa et al., *Robot Path Planning and Cooperation*, Studies in Computational Intelligence 772, https://doi.org/10.1007/978-3-319-77042-0_3

requirement in mobile robot navigation. In fact, we can tolerate some deviation with respect to optimality for the sake of reducing the execution time of the path planning algorithm, since in real robotics applications, it does not harm to generate paths with slightly higher lengths, if they can be generated much faster. We also design new hybrid algorithms combining the different algorithms with the intention to improve the performance of the designed algorithms and explore the benefit of hybridization on global robot path planning.

3.2 System Model

The workspace of the mobile robot is represented by a grid-based map. The grid map is constituted from a number of cells of equal sizes. Each cell is identified by unique number, beginning from 0 for the top left cell, 1 for the next cell to the right, and so on. Figure 3.1 shows an example of 10×10 grid map. The reasoning behind choosing the grid map model is the integration of our planners in the Robot Operating System (ROS) framework [2] for robot application development. ROS uses two-dimensional matrix of cells called "occupancy grid map" to represent the grid map. Each cell, in the occupancy grid map, represents a certain square area of the environment which stores the occupancy information (occupied or not by an obstacle). A unique number represents the occupancy information which can have one of three possible values 0, 100, or -1. '0' means the cell is free, '100' means the cell is occupied by an obstacle, and '-1 'means the cell is unknown. Figure 3.1 shows an example of a 10×10 occupancy grid map. We suppose that the robot movement from one cell to another is in horizontal, vertical, and diagonal directions (8 adjacent grid cells at most). The obstacles are static and known in advance.

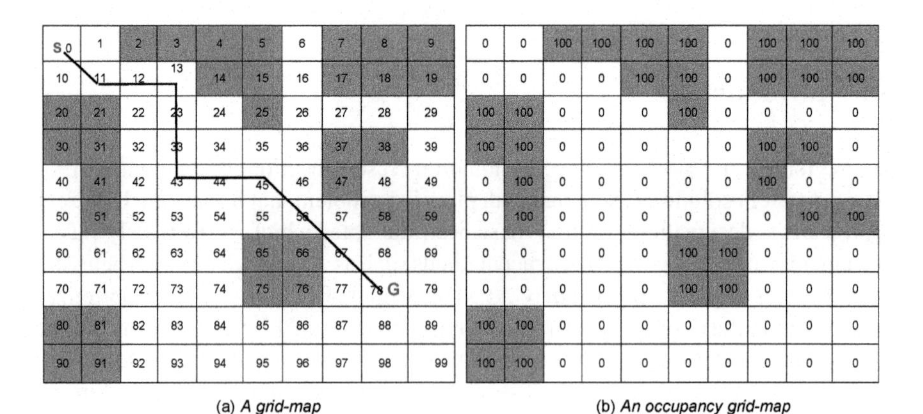

Fig. 3.1 A 10×10 grid environment

The robot's path is constituted from a series of free cells beginning with the start cell to the goal cell. Figure 3.1 shows an example of a feasible path from the start cell 0 to the goal cell 78 (0, 11, 12, 13, 23, 33, 43, 44, 45, 56, 67, 78). The path cost is the sum of the robot movement sequence. The vertical and horizontal moves in the grid are calculated as 1 unit distance, whereas the diagonal moves are calculated as 1.4 unit distance. Thus, the path cost is (1.4 + 1 + 1 + 1 + 1 + 1 + 1 + 1 + 1 + 1.4 + 1.4 + 1.4) = 13.6. During the search process, numerous paths are generated that may have variable path lengths or sizes.

3.3 Design of Exact and Heuristic Algorithms

We have proposed and designed carefully the iPath library [3] that provides the implementation of several path planners according to the following two classes of methods: (1) heuristic method and (2) graph searching or exact algorithms. The $iPath$ library is available as open source on Github under the GNU GPL v3 license. The documentation of API is available in [4]. We designed six different algorithms: the A^* algorithm, a relaxed version of A^* at the aim to ameliorate the performance of A^*, the Tabu Search algorithm, the genetic algorithm, the neural network algorithm, and the ant colony optimization algorithm. Indeed, looking at the literature, there exist several attempts to compare or combine two or three approaches together. In our literature review, we noticed that some research works have presented a solution for the path planning problem based on a combination of two or three approaches. To the best of our knowledge, there is no major research efforts that presented a comprehensive comparative study between the most used approaches in the literature for mobile robot global path planning in grid environments. This work aims at filling this gap by comparing six different approaches. This section discusses the basic principles and concepts pertaining to applying the different algorithms for solving the global path planning problem in grid environments.

3.3.1 A Relaxed Version of A for Robot Path Planning*

The classical A^* algorithm may be time-consuming to reach the optimal solution for hard instances (such as mazes) depending on the density of obstacles [5]. In order to overcome these drawbacks, we propose a novel algorithm called relaxed AStar (RA^*) [6]. The core idea consists in exploiting the grid map structure, typically used as the environment model for the global path planning problem in mobile robotics, and we establish an accurate approximation of the cost of the optimal path between two cells. Namely, we provide lower and upper bounds of the cost of the optimal path and we show that this cost could be approximated in terms of the number of moves on that path. Based on that approximation, we designed a linear relaxed variant RA^* of A^* technique.

In Chap. 2, we presented some research works that have tried to improve the performance of A^*. The main difference of these algorithms with our variant RA^* is that the relaxing is carried out on the exact cost g of the evaluation function f ($f = g + h$) instead of the heuristic h as usually done in the existing relaxations of A^*. The amount g is computed at most only once for every cell during the entire search; it is determined when a node is visited the first time and then $f = g + h$ remains constant during the rest of the search. The advantage of our approaches is the saving of the processing time to find the (near)-optimal solution.

3.3.1.1 Approximation of the Optimal Path

Let G4-grid and G8-grid denote a regular grid where each cell has at most four and eight neighbors, respectively. Precisely, a path in a G4-grid allows only horizontal and vertical moves, while a path in a G8-grid allows in addition diagonal moves. In a regular grid map, the cost of a move is defined to be equal to one unit for horizontal and vertical moves. For diagonal moves, the cost is equal to $\sqrt{2}$.

Let P be a path from a start cell to a goal cell, $|P|$ denotes the number of moves on P, and $cost\ (P)$ denotes the sum of move costs of P. The following property provides bounds of a given path P in a regular grid.

Property 1:
For any path P in a regular grid map, we have:
$|P| \leq cost\ (P) \leq \sqrt{2}\ |P|$

Proof:
Let x be the number of horizontal or vertical moves in P and let y be the number of diagonal moves, then we have:
$|P| = x+y$.
$cost\ (P) = x + \sqrt{2}\ y$
We deduce that:
$|P| = x+y \leq cost\ (P) = x+ \sqrt{2}y \leq \sqrt{2}x+ \sqrt{2}y=\sqrt{2}|P|$
Thus, it is clear that the property holds.

In G4-grid, we establish the following stronger property:

Property 2:
For any path P in a G4-grid, $cost\ (P) = |P|$.

Proof:
Using the same notations as in the proof of property 1, since in a G4-grid $y = 0$ then $cost\ (P) = |P| = x$.
Let P(S,G) be the set of feasible (obstacle-free) paths from cell S to cell G. As a direct consequence of property 2, in G4-grid we have: $\mathrm{argmin}_{P \in P(S,G)}|P|$ is the optimal path from S to G, while this statement is not necessarily true in G8-grids. For instance in the example below: $\mathrm{argmin}_{P \in P(S,G)}|P| = (S,A,B,C,G)$. For this path, $|(S,A,B,C,G)| = 4$, and $cost(S,A,B,C,G) = 4\sqrt{2} = 5.6$, while there exists a path (S,D,E,F,I,G)

that has greater number of moves ($|(S,D,E,F,I,G)|$=5) but a better cost equal to 1+1+1+1+$\sqrt{2}$ = 5.4. The gap between the optimal path (S,D,E,F,I,G) and the path (S,A,B,C,G) which has the minimum number of moves is 0.2 (+3.7%). Thus, our proposed method favors the number of moves as the cost metric over the actual distance (i.e., unit cost in all 8 directions). In the extreme case, where all moves are diagonal, there might be a $\sqrt(2) = 1.41$ ratio between the solution found by the proposed methods and the actual shortest path found using the horizontal and vertical edges on the graph. The simulation results will show that the gap is much smaller in practice and that in most cases the path with the smallest number of moves is actually the shortest path even in G8-grids.

H	I	G	
F			C
E		B	
D	A		
S			

3.3.1.2 The *RA** algorithm

A* algorithm expands the nodes of the tree-search according to the evaluation function $f(n) = g(n) + h(n)$, where $g(n)$ is the exact cost of the path from the root (start node) to the node n of the tree-search and $h(n)$ is a heuristic function estimating the remaining path to the goal. Hence, $f(n)$ is an estimation of the optimal path from the root to a goal traversing the node n. It is proven that A* finds the optimal solution when the heuristic h is consistent [7]. Notice that the exact cost $g(n)$ of a node n may be computed many times; namely, it is computed for each path reaching node n from the root. Based on *Property 2*, the entity $g(n)$ may be approximated by the cost of the minimum-move path from the start cell to the cell associated with node n. The idea of the RA* algorithm is to approximate the exact costs of all paths leading to node n from the root of the tree-search by the exact cost of the first path reaching n. Thus, unlike standard A* algorithm, $g(n)$ is computed only once since the optimal path reaching node n from the start cell S is approximated by $\text{argmin}_{P \in P(S,n)} |P|$. Since both terms $g(n)$ and $h(n)$ of the evaluation function of the RA* algorithm are not exact, then there is no guaranty to find an optimal solution. *Algorithm. 1* presents the standard A* pseudocode. In order to obtain the relaxed version RA*, we simply skip some instructions (highlighted in the algorithm above) that are time-consuming with relatively low gain in terms of solution quality.

The main idea in the relaxed version is that a node is never processed more than once. Consequently, we no longer need a closed set (lines 1, 13, and 15). Besides, in order to save time and memory, we do not use a *came_from* map (lines 3 and 20). Instead, after reaching the goal, the path can be reconstructed, from goal to start by

Algorithm 6. The relaxed Astar Algorithm RA^*

```
   input : Grid, Start, Goal
 1 tBreak = 1+1/(length(Grid)+width(Grid));
   // Initialisation:
 2 openSet = Start // Set of nodes to be evaluated;
 3 for each vertex v in Grid do
 4 |  g_score(v)= infinity;
 5 end
 6 g_score[Start] = 0;
   // Estimated total cost from Start to Goal:
 7 f_score[Start] = heuristic_cost(Start, Goal);
 8 while openSet is not empty and g_score[Goal]== infinity do
 9 |  current = the node in openSet having the lowest f_score;
10 |  remove current from openSet;
11 |  for each free neighbor v of current do
12 |  |  if g_score(v) == infinity then
13 |  |  |  g_score[v] = g_score[current] + dist_edge(current, v);
14 |  |  |  f_score[v] = g_score[v] + tBreak * heuristic_cost(v, Goal);
15 |  |  |  add neighbor to openSet;
16 |  |  end
17 |  end
18 end
19 if g_score(goal) ! = infinity then
20 |  return reconstruct_path(g_score) // path will be
   |  reconstructed based on g_score values;
21 else
22 |  return failure;
23 end
```

selecting, at each step, the neighbor having the minimum g_score value. Moreover, we do not need to compare a tentative g_score to the current g_score (line 19) since the first calculated g_score is considered definite. This is an approximation that can be correct or not. Simulations will later assess the quality of this approximation. Finally, we do not need to check whether the node is in the open list (line 23). In fact, if its g_score value is infinite, it means that it has not been processed yet, and hence is not in the open list. The tie-breaker factor (slightly > 1) which multiplies the heuristic value (line 4) is a well-known technique that is used so that the algorithm favors a certain direction in case of ties. If we do not use tie-breaking (tBreak=1), the algorithm would explore all equally likely paths at the same time, which can be very costly, especially when dealing with a grid environment, as will be confirmed by the simulations. Figure 3.2 shows a simple example of several equivalent optimal paths between two nodes in a G4-grid. Actually, in this example, any obstacle-free path containing 3 moves up and 10 moves right, in whatever order, is an optimal path. A similar example can be shown in a G8-grid. If we do not use a tie-breaking factor, A* would explore all equivalent paths at the same time. In fact, at a given iteration, several nodes belonging to different paths would have the same f_score, which means that, in the following iterations, each time one of these nodes would be chosen randomly with no preference given to any of these paths, while with the tie-breaker factor, when two nodes have the same value of the sum ($g\_score + h\_score$), the node having a minimum h_score (presumably closest to the goal) would be

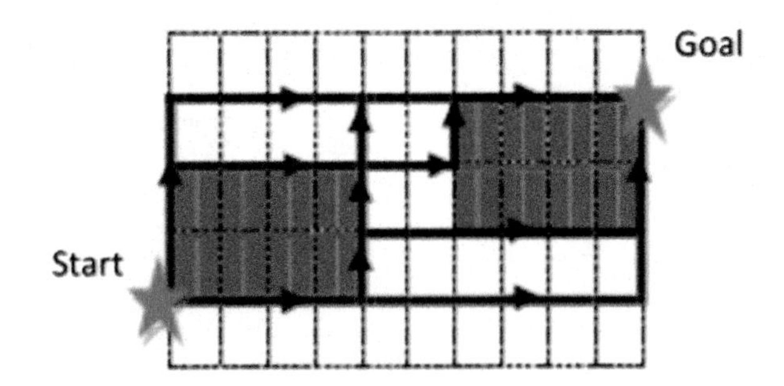

Fig. 3.2 Example of several equivalent optimal paths between two nodes in a G4-grid. Obstacles are in gray

expanded first. Thus, the algorithm will follow only one of the paths until it either hits the goal or finds an obstacle. We chose tBreak$= 1 + \epsilon$, with ϵ a small positive equal to 1/(mapWidth + mapLength). This assures that the new heuristic function $(1+\epsilon)$*h remains consistent. In fact, whether h is the Manhattan distance or the shortcut distance, it is bounded by (mapWidth + mapLength), which means that ϵ is always smaller than 1/h . Consequently, $(1 + \epsilon)$*h $\leq$ h + 1, which means that the solution cost is guaranteed to be strictly less than optimal cost+1. The tie-breaking technique has already been used in variants of A* algorithm [8]. RA* pseudocode is presented in *Algorithm 6*. Highlighted instructions are specific to the relaxed version. Instead of having a closed set, nodes having a finite g_score value are considered as processed and closed (lines 3–5, 8, 12, 19). And after reaching the goal, and quitting the main loop, the path is reconstructed (line 20), from goal to start, based on g_score values (at each step, we add to the head of the path the neighbor node presenting the minimum g_score value).

3.3.2 The Tabu Search Algorithm for Robot Path Planning (TS-PATH)

In this paragraph, we present the Tabu Search algorithm TS-PATH. For more details, the reader can refer to [9]. The pseudocode of the algorithm is presented in Algorithm 7.

1. **Generation of the initial solution**: The TS-PATH algorithms begin with the generation of an initial solution or path from the source position to the destination position. The greedy method based on the euclidian distance heuristic is used to construct an initial feasible path. The neighbor cells forming the path must be connected and do not contain obstacles. During the path search process, a robot can fall in a deadlock position. In such situation, the robot is blocked and can not evolve toward the goal as it is surrounded by obstacles or already visited positions.

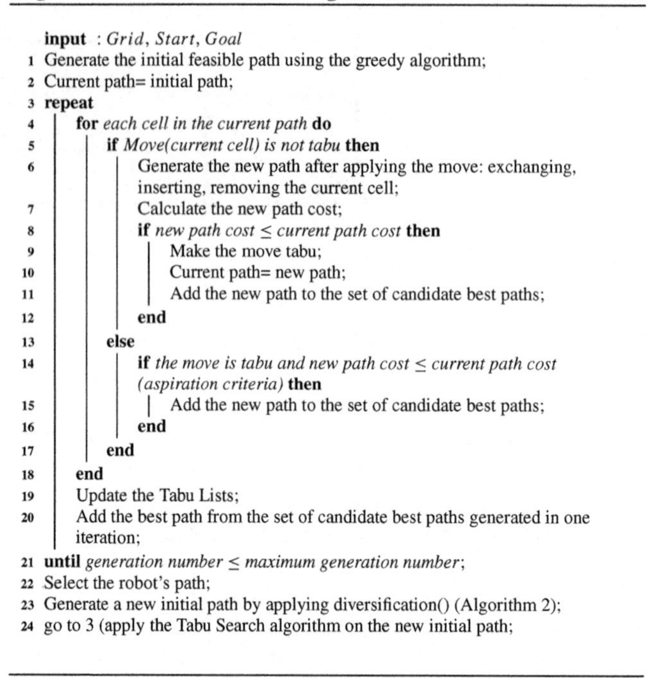

Algorithm 7. The Tabu Search Algorithm $TS - PATH$

input : *Grid, Start, Goal*

1 Generate the initial feasible path using the greedy algorithm;
2 Current path= initial path;
3 **repeat**
4 **for** *each cell in the current path* **do**
5   **if** *Move(current cell) is not tabu* **then**
6   Generate the new path after applying the move: exchanging, inserting, removing the current cell;
7   Calculate the new path cost;
8   **if** *new path cost* $\leq$ *current path cost* **then**
9   Make the move tabu;
10   Current path= new path;
11   Add the new path to the set of candidate best paths;
12   **end**
13   **else**
14   **if** *the move is tabu and new path cost* $\leq$ *current path cost (aspiration criteria)* **then**
15   Add the new path to the set of candidate best paths;
16   **end**
17   **end**
18 **end**
19 Update the Tabu Lists;
20 Add the best path from the set of candidate best paths generated in one iteration;
21 **until** *generation number* $\leq$ *maximum generation number*;
22 Select the robot's path;
23 Generate a new initial path by applying diversification() (Algorithm 2);
24 go to 3 (apply the Tabu Search algorithm on the new initial path;

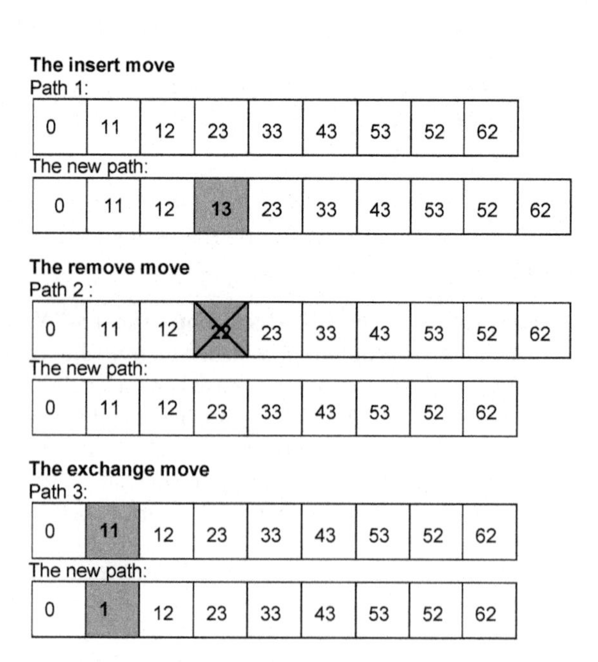

Fig. 3.3 Insert, remove and exchange moves

To recover from this situation, a backtracking operation is applied which enables the robot to move back to an earlier visited cell surrounded by free unvisited cells.

2. **Current solution neighborhood generation**:

- *The Definition of the Neighborhood Structure*: The initial path constructed in the first step constitutes the entry point for the TS-PATH algorithm. As previously cited in Chap. 2, Tabu Search metaheuristic is a local search method, meaning that it tries, in each iteration, to ameliorate the current solution and create the neighborhood by applying a transformation called move. This process is repeated until a predefined condition is satisfied. For TS-PATH algorithm, we designed three different moves: (1) insert move, (2) remove move, and (3) exchange move. A new feasible path must be created after applying one of these transformations to the current feasible path.
  Consider any path p of length L from the source position S to the destination position G. The neighborhood of p is obtained by applying the following moves:
  - Cell removing: Verify the possibility to remove each cell of p different from S and G in order to generate a new feasible path $p\prime$.
  - Cell exchanging: Verify the possibility to replace each cell c of p, different from S and G, by another cell $c\prime$ not in p in order to generate a new feasible path $p\prime$.
  - Cell insertion: Verify the possibility for each cell c not in p to insert it to p in order to generate a new feasible path $p\prime$.

  Figure 3.3 shows the three described moves. In the first example, cell 13 of path 1 is inserted between cell 12 and cell 23. In the second example, cell 22 is removed from path 2. In the third example, cell 11 in path 3 is exchanged by cell 1. With respect to the map in Fig. 3.1, the new paths are feasible.

- *Tabu List, tenure*: As mentioned previously in Chap. 2, the Tabu List mechanism is used in the Tabu Search approach to prevent cycling back to previously visited solutions. We devised two Tabu in the TS-PATH algorithm: TabuListIn and TabuListOut. These lists are of maximum size 2 x tenure. After performing a move, some arcs are added to the current path and these arcs are added to TabuListIn. Other arcs are removed from the current path, and these arcs are added to TabuListOut. Four attributes characterize each move (cell i, cell j) in a list: "fromCell," "ToCell," "tenure," and "ExpirationDate." In the following example, cell 11 and cell 21 are exchanged. The arcs (10,21) and (21,20) are inserted in TabuListIn, and the arcs (10,11) and (11,20) are inserted in tabuListOut. The different attributes of the arc (10,21): fromCell = 10, toCell = 21, tenure = k, ExpirationDate = current number of generation + tenure. This is to avoid the algorithm from cycling back on this move during (number of generation + tenure) iterations. A tabu status is assigned to the arc stored in the Tabu List. As mentioned previously, the maximum size of a Tabu List is 2 x tenure. In fact, after performing one move the maximum number of arcs that can be added to the TabuListIn or to TabuListOut is 2. Moreover, each move stays tabu in the Tabu List stays "tenure" iterations. Thus, the maximum size of one list will not exceed 2 * tenure. Before carrying out a move, we must check that the move is not tabu; i.e., the arcs that will be inserted in the current path do not exist in TabuListIn and the arcs that will be removed from the current path do not exist in the TabuListOut. The update of the Tabu Lists is done at the end of each iteration of TS-PATH. If the expirationDate of one move is equal to the current iteration number, then the move will be not tabu and it will be removed from the list.

- *Aspiration Criteria*: During the current path neighborhood generation, some moves that are tabu and that are able to improve the best current solution quality could be accepted and liberated from the Tabu List. This method is a way to relax the mechanism of the Tabu List. It is called aspiration criteria. Before accepting that move, we must verify two conditions: First, we must be sure that it will ameliorate the best current path. Second, the algorithm will not go back to the already explored paths.

3. **Step 3: Incremental cost evaluation of neighbor paths**: Each neighbor path cost must be quickly evaluated. We adopted an incremental computing of the following new paths' costs as follows: Let p' be a feasible neighbor path of the path p, obtained by applying the three different moves mentioned above (insert, remove, exchange).

   - After insertion of cell x between two cells $x1$ and $x2$ of p, then $cost(p') = cost(p) - cost(x1, x2) + cost(x1, x) + cost(x, x2)$.
   - After removing cell x from p between $x1$ and $x2$, then $cost(p') = cost(p) + cost(x1, x2) - cost(x1, x) - cost(x, x2)$.
   - After exchanging cell x in p between cells $x1$ and $x2$ by cell y not in p, then $cost(p') = cost(p) - cost(x1, x) - cost(x, x2) + cost(x1, y) + cost(y, x2)$.

   The incremental computing of the costs leads to save time and permits to explore large neighborhoods.

4. **Step 4: Diversification**:
   After a certain number of iterations, the search could be stagnated. In such case, the diversification mechanism is used to drive the search toward a new region of the search space. The first step of the diversification method consists in drawing a straight line between the source and the destination positions using the greedy method based on the Euclidean distance heuristic. This line could contain some obstacles. At the radius of N cells (N is a random parameter), the algorithm chooses a random intermediate cell, which will be used to generate a new feasible path from the start to the goal cells across it. The new generated path will be used as an initial solution to restart the Tabu Search. The fact that the new path involves an intermediate cell generated randomly gives a chance to explore

different regions than those explored so far. The pseudocode of diversification is presented in Algorithm 8.

Algorithm 8. Diversification

1 Draw the straight line from the initial position to the goal position. ;
2 $Center_{Cell}$: Select randomly a point near the center cell of the straight line.;
3 $N_{RadiusN}$: set of $Center_{Cell}$ neighbors at Radius N, N is a parameter. ;
4 $Cross_{Point}$: Select randomly one cell from $N_{RadiusN}$.;
5 $Path_{S,CP}$: find the path using the greedy approach between initial position and $Cross_{Point}$. ;
6 $Path_{CP,G}$: find the path using the greedy approach between $Cross_{Point}$ and the goal position. ;
7 $Path_{Robot}$= $Path_{S,CP} \cup Path_{CP,G}$.

3.3.3 The Genetic Algorithm for Robot Path Planning

This section presents the proposed genetic algorithm [10, 11]. The proposed algorithm is presented in Algorithm 9. Following is the description of each step of the proposed GA algorithm in detail.

1. *Initial Population*: The proposed GA algorithm begins with generating a set of feasible candidate paths to form the initial population. Thus, each individual in the population should be continuous and not across any obstacles. Starting with only feasible solutions has advantages on improving the execution time and the solution quality [12, 13]. This process starts with constructing initial path using the Greedy approach (with Euclidean distance heuristic) from the start cell to the goal cell. The Greedy approach is used to construct the paths in short times. Then to generate the next individuals, the algorithm will choose randomly some intermediate cell not in the initial path, then it will use the greedy approach to connect the start cell to this intermediate cell, and then connect it with the goal cell. Choosing efficiently the initial population is one of the common problems in the GA which affects directly the performance of the algorithm. Thus to give the GA the ability to better explore the area, the algorithm will choose far intermediate cells in large environments.
2. *Fitness Evaluation*: The fitness evaluation of the different candidates starts when the initial population achieves its maximum size.
   Each individual in each generation will be evaluated using a fitness function. The fitness function will assign to every individual a value which represents its goodness based on some metrics. In the proposed approach, the fitness function

Algorithm 9. The genetic algorithm

```
   input : Grid, Start, Goal
   output: dist
 1 Generate the initial population (using the greedy approach);
 2 while (Number of generation < Maximum number of generation) do
 3     Perform fitness evaluation for each individual;
 4     Perform the elitist selection;
 5     Perform the Rank selection;
 6     repeat
 7         From current generation choose two parents randomly;
 8         if (random number generated < the crossover probability) then
 9             if (the parents have common cells) then
10                 perform the crossover;
11                 move the resultant paths to the next generation;
12             else
13                 choose other two parents;
14             end
15         else
16             move the parents to the next generation;
17         end
18     until (Population size of next generation = Maximum size of population);
19     for each individual in the next generation do
20         if (random number generated < the mutation probability) then
21             choose one cell randomly;
22             replace the cell with one of its neighbors ;
23             if (the resultant path is feasible) then
24                 replace old individual with the new individual ;
25             else
26                 choose two cells C1 and C2 randomly from the path;
27                 remove all the cells between C1 and C2;
28                 use the greedy approach to connect C1 and C2;
29             end
30         end
31     end
32 end
```

used the *path length* as a primary metric for the evaluation process. In the path planning problem, the fitness of a path should increase when the length of the path decreases; therefore, the fitness function in Eq. (3.1) is used:

$$F = \frac{1}{\sum_{i=1}^{n} d_{i,i+1}} \tag{3.1}$$

where $d_{i,i+1}$ is the Euclidean distance between cell i and cell $i+1$ and n is the number of cells forming the path.

3. *Selection*: The selection operator simulates the survival of the fittest principle, meaning that it gives preference to the better solutions by allowing them to pass to the next generation. It will be conducted after the evaluation to select the best individuals from the current generation. In the literature, many selection strategies were introduced such as elitist selection, rank selection, tournament selection, and roulette wheel selection. In the proposed algorithm, both the elitist selection and the rank selection mechanisms were used. First, the elitist selection is applied to

keep the best individual from the population and move it to the next generation without any change. It is used to prevent losing good solutions because of the randomness of the GA operators. Next, the rank selection will be performed and it starts with sorting the population in ascending order based on the path length. Then, each individual will be assigned with a probability value based on its length such that the shorter path will receive higher probability value. The rank selection is opted to be used because it increases the diversity as each individual will has a chance to be selected.

Now, the current generation is formed from the individuals that are shown to be good temporary. The next step consists in applying the different genetic operators to the current generation. These operators are crossover and mutation. They aim at creating a new generation of individuals and hopefully better individuals.

4. *Crossover*: The crossover operator is the primary operator in GA. It is a probabilistic mechanism, and the GA relies on it to explore the search space by exchanging information between the solutions. It is also called recombination because it simulates the gene recombination process by choosing two possible solutions (called parents) and then trying to recombine their genes to hopefully get enhanced solutions from them (called offspring).
   In the literature, many crossover strategies were introduced such as one-point crossover, two-point crossover, and uniform crossover.
   Three different crossover strategies were implemented aiming at comparing their performance: one-point crossover, two-point crossover, and a modified crossover strategy that proposed in [14]. Figure 3.4 shows an example for the three methods. The crossover starts by choosing randomly two parents from the population. In the one-point crossover, one common cell between the parents will be selected randomly. Then, the parents will exchange the parts after that common cell to produce two offspring. In the two-point crossover, two common cells between the parents will be selected randomly. Then, the parents will exchange the parts between those common cells to produce two offspring. In the modified crossover, two common cells between the parents will be selected randomly. Then, the best (shorter) part from each parent is chosen to produce one offspring as illustrated in Fig. 3.4.
5. *Mutation*: The second genetic operator is the mutation. Each individual in the population is subjected to the mutation operator which provides the diversity into the population. Although the mutation is considered a secondary operator in GA, it ultimately helps in escaping from the local optimum situations. It aims at adding new information to the genetic search process in a random way.
   In the proposed algorithm, the mutation will be performed after the crossover. When the individuals is selected for mutation, the algorithm will choose randomly one of the path genes to replace it with one of that gene neighbors. If the resultant path from that replacement is feasible, the new path will be accepted immediately. Otherwise, if after several trials, the resultant path is infeasible, the algorithm will choose randomly two genes G1 and G2 from the path, all the genes between them will be removed, and then G1 and G2 will be reconnected by constructing new path between them using the Greedy approach. Old neighbor of G1 in the path

will be discarded from the search to guarantee that the new path between G1 and G2 is different from the old one.

6. *Termination Condition*: The algorithm is iterated until a maximum number of generations is reached. The chromosomes evolve from one generation to the other. After several generations, the fittest paths survive and an (or near) optimal solution is obtained.
7. *Control Parameters*: The GA performance is affected by a number of parameters: number of generations, population size, crossover probability, and mutation probability. The number of generations is usually used as a stopping criterion to determine the number of iterations after which the GA should stop. However, other stopping criteria could be used. The GA could be terminated when an individual has some certain fitness value or when all individuals in the population have a certain degree of similarity.
   Together, population size, crossover probability, and mutation probability are called *control parameters*. For GA search to work properly, it must maintain a balance between its two main functionalities:

   - Explore new areas from the search space.
   - Exploit the already explored areas.

   So, maintaining the balance in combining those tasks critically controls the GA performance. This balance is determined by wisely adjusting the three control parameters.
   The population size specifies the number of individuals in each population. Larger population size means that the GA could explore larger areas from the environment, which will decrease the possibility of falling into local optimum. As a consequence, larger population size means longer execution time. The probabilities (or rates) of the genetic operators (i.e., crossover and mutation) specify how frequently they should be performed. After determining the individuals for the operator, the algorithm will only perform it if a uniformly distributed randomly generated number in the range [0,1] is less than the operator probability (i.e., crossover probability, mutation probability). Otherwise, the individuals will remain without changes and moved to the next generation. Higher crossover probability will make the process of generating new individuals faster, but at the same time it could increase the disruption of the good solutions. Higher mutation probability will lead to enhance the diversity of population, but at the same time tends to transform the GA into a random search. Typical crossover probability will be in ranges [0.6 to 0.9], and typical mutation probability will be in range [0.001,0.01] [15].

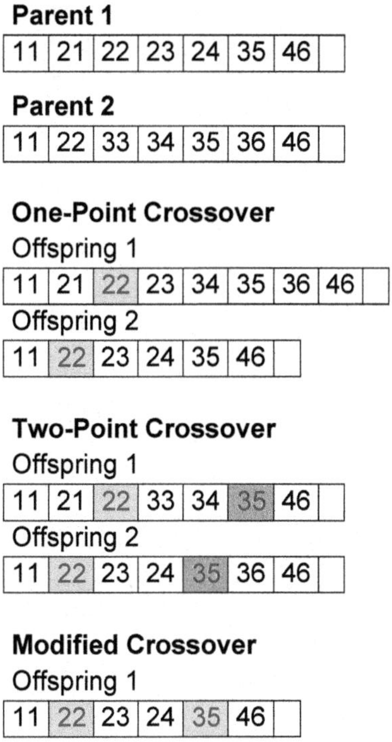

Fig. 3.4 Crossover operators

Parent 1

11	21	22	23	24	35	46	

Parent 2

11	22	33	34	35	36	46	

One-Point Crossover

Offspring 1

11	21	22	23	34	35	36	46	

Offspring 2

11	22	23	24	35	46	

Two-Point Crossover

Offspring 1

11	21	22	33	34	35	46	

Offspring 2

11	22	23	24	35	36	46	

Modified Crossover

Offspring 1

11	22	23	24	35	46	

3.3.4 The Ant Colony Optimization Algorithm for Robot Path Planning

The pseudocode of the ACO algorithm is presented in Algorithm 10. In what follows, we describe the different steps of the algorithm:

1. *Step 1*: *Initialization*: Suppose Ants = $\{Ant_1, Ant_2, ..., Ant_m\}$ denotes a set of m ants. Let $\tau_{i,j}(t)$ be the intensity of pheromone on edge (i,j) at time t.
   At time zero, an initialization phase takes place during which the m ants are positioned on the start position and the initial value of pheromone $\tau_{i,j}(0)$ is set to a small positive constant c. There are also a lot of parameters in ACO algorithm that need to be fixed.
   - α: is the pheromone factor;
   - β: is the heuristic factor;
   - ρ: is the evaporation rate;
   - Number of iterations;
   - Q: is a constant.

   In order to forbid that an ant visits a sensor node more than once, we associate with each ant a **Tabu List** that saves the sensor nodes already visited.

2. *Step 2*: *Path finding*: The ants search for the shortest path in the environment from the start to the goal positions. At each iteration, each ant has a current position and has to smartly decide the next cell on its path toward the destination. The next cell is chosen according to the *Transition Rule Probability*

Algorithm 10. The Ant Colony Algorithm

```
   input : Grid, Start, Goal
   output: dist
1  for each ant do
2  |  Add the start position to the ants paths;
3  end
4  repeat
5  |  for each ant do
6  |  |  if the ant is not blocked in a dead-end position then
7  |  |  |  Choose the next position according to equation ...;
8  |  |  |  Add the next position to the ant's path;
9  |  |  end
10 |  |  Add the ant's path to the set of best iteration i;
11 |  end
12 |  Generate the best path found in iteration i;
13 |  Update pheromone according to equation ;
14 |  Add the best path of iteration i to the set of best paths;
15 until (Number of iteration i = Maximum number of ieration);
16 Generate the robot path from the set of best paths
```

$$p_{i,j}^{k} = \left\{ \begin{array}{cc} \frac{\tau_{i,j}^{\alpha} * \eta^{\beta}}{\sum_{j \in allowed(i)} (\tau_{i,j}^{\alpha} * \eta^{\beta})} & \text{if } j \in allowed(i) \\ 0 & \text{otherwise} \end{array} \right\} \tag{3.2}$$

where $\tau_{i,j}$ denotes the quantity of pheromone between cell i and cell j, $allowed\ (i)$ is the set of neighboring nodes of cell i, which the kth ant has not visited yet, α is the pheromone factor, β is the heuristic factor, and we call visibility η the quantity $\frac{1}{d_{i,j}}$, where $d_{i,j}$ is the distance between the current cell i and the next cell j.

3. *Step 3*: *Pheromone update*: At the beginning of the algorithm, an initial pheromone value is affected to the edges of the grid. After each iteration of the algorithm, the quantity of pheromone is updated by all the ants that have built solutions. The quantity of pheromone $\tau_{i,j}$, associated with each edge joining two vertices i and j, is updated as follows:

$$\tau_{i,j}(t+1) = (1-\rho) * \tau_{i,j}(t) + \sum_{k=1}^{m} \Delta\tau_{i,j}^{k}(t) \tag{3.3}$$

where $0 \leq \rho \leq 1$ is the evaporation rate, m is the number of ants, and $\Delta\tau_{i,j}^{k}$ is the quantity of pheromone laid on edge (i, j) joining two positions i and j by an ant k.

$$\Delta\tau_{i,j}^{k}(t) = \left\{ \begin{array}{ll} \frac{Q}{L_k} & \text{if ant k used edge(i,j) in its tour} \\ 0 & \text{otherwise} \end{array} \right\} \tag{3.4}$$

where Q is a constant and L_k is the length of the tour constructed by ant k.

3.4 Performance Analysis of Global Path Planning Techniques

3.4.1 Simulation Environment

To perform the simulations, we implemented the iPath C++ library [3] that provides the implementation of several path planners including A^*, GA, Tabu Search, Ant colony optimization, and RA^*. The iPath library is implemented using C++ under Linux OS. It is available as open source on Github under the GNU GPL v3 license. The documentation of API is available in [4]. A tutorial on how to use iPath simulator [Online]. Available: http://www.iroboapp.org/index.php?title=IPath .

The library was extensively tested under different maps including those provided in the benchmark defined by [16] and other randomly generated maps [17]. The benchmark used for testing the algorithms consists of four categories of maps:

1. **Maps with randomly generated rectangular-shape obstacles**: This category contains two (100 × 100) maps, one (500 × 500) map, one (1000 × 1000) map, and one (2000 × 2000) map with different obstacle ratios (from 0.1 to 0.4) and obstacle sizes (ranging from 2 to 50 grid cells).
2. **Mazes**: All maps in this set are of size 512 × 512. We used two maps that differ by the corridor size in the passages (1 and 32).
3. **Random**: We used two maps of size 512 × 512.
4. **Rooms**: We used two maps in this category, with different room sizes (8 * 8 and 64 * 64). All maps in this set are of size 512 × 512.
5. **Video games**: We used one map of size 512 * 512.

The simulation was conducted on a laptop equipped with an Intel core i7 processor and an 8GB RAM. For each map, we conducted 5 runs with 10 different start and goal cells randomly fixed. This makes 600 (12 maps x 5 × 10) total runs for each algorithm (Fig. 3.5 and Table 3.1).

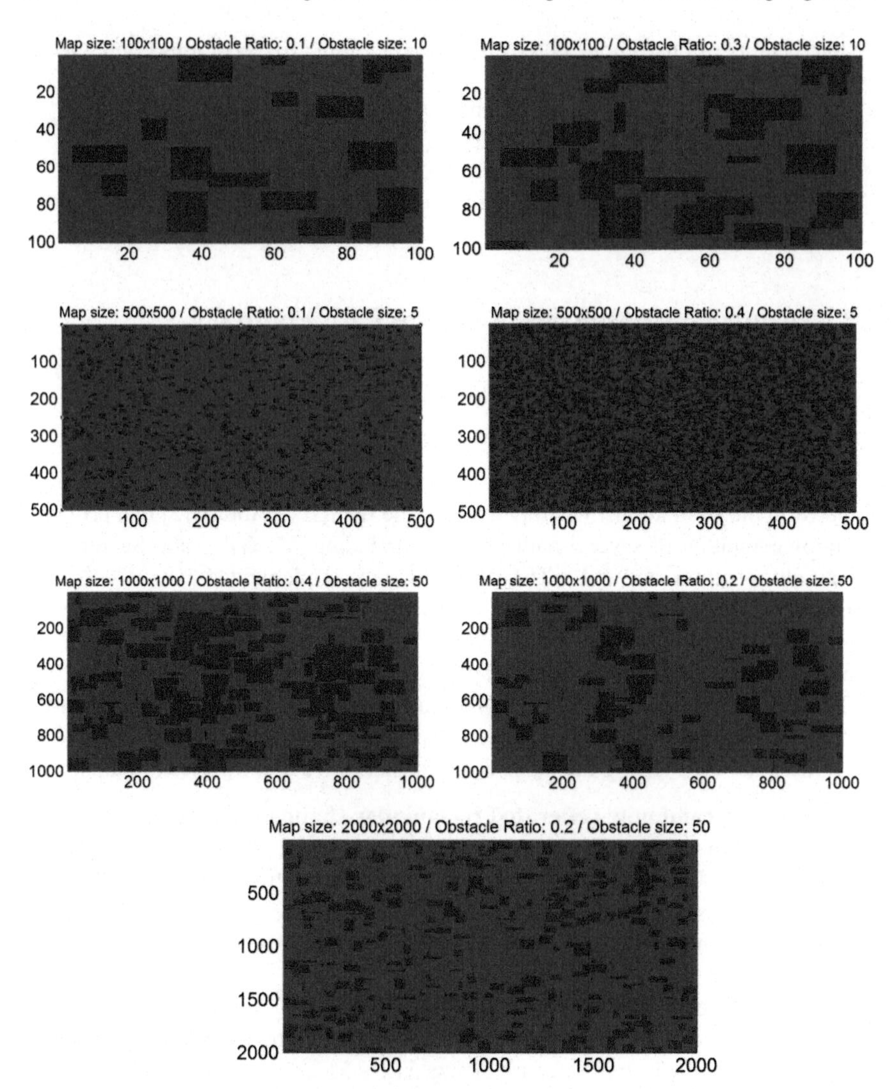

Fig. 3.5 Examples of maps used for the simulation

Table 3.1 Average path cost (grid units) for the different algorithms, per environments size

Algorithm	100×100	500×500	1000×1000	2000×2000	Random (512×512)	Rooms (512×512)	Video games (512×512)	Mazes (512×512)
A^*	66.2	273.8	473.4	1366.4	303.4	310.6	243.6	1661.9
RA^*	70.5	277.0	477.7	1443.3	319.5	341.8	257.8	1687.1
GA	75.3	293.3	476.8	–	360.8	408.7	269.9	–
TS	69.6	335.9	541.1	–	322.9	322.9	531.7	–
NN	82.3	282.8	487.2	–	485.4	322.9	253.1	–

3.4.2 Simulation Results

In this section, we present the simulation results relative to the evaluation of the efficiency of the five different planners. Figures 3.6 and 3.7 depict the box plot of the average path costs and average execution times for the randomly generated maps and the benchmarking maps, respectively, with different start and goal cells. Figure 3.8 shows the average path cost and the average execution time of all the maps. On each box, the central mark is the median and the edges of the box are the 25th and 75th percentiles. Tables 5.2 and 3.2 present the average path costs and average execution times for the different maps. Table 3.3 presents the percentage of extra lengths of the different algorithms in different types of maps, and Table 3.4 presents the percentage of optimal paths found by the algorithms. We can conclude from these figures that the algorithms based on heuristic methods are in general not appropriate for the grid path planning problem. In fact, we observe that these methods are not as effective as RA^* and A^* for solving the path planning problem, since the latter always exhibit the best solution qualities and the shortest execution times. Although GA can find optimal paths in some cases as shown in Table 3.4, it exhibits higher runtime as compared to A^* to find its best solution. Moreover, non-optimal solutions have large gap 15.86% of extra length on average as depicted in Fig. 3.9. This can be explained by two reasons: The first reason is that GA needs to generate several initial paths with the greedy approach and this operation itself takes time which is comparable to the execution of the whole A* algorithm. The second reason is that GA needs several iterations to converge and this number of iterations depends on the complexity of the map and the positions of the start and goal cells.

The Tabu Search approach was found to be the least effective. It finds non-optimal solutions in most cases with large gaps (32.06% as depicted in Fig. 3.9). This is explained by the fact that the exploration space is huge for large instances, and the Tabu Search algorithm only explores the neighborhood of the initial solution initially generated. On the contrary, the neural network path planner could find better solution qualities as compared to GA path planner and Tabu Search path planner. We can see from Fig. 3.9 that the average percentage of extra length of non-optimal paths found by this planner does not exceed 4.51% . For very large (2000×2000) and complex grid maps (maze maps), heuristic algorithms fail to find a path as shown in Tables 5.2 and 3.2. This is due to the greedy approach which is used to generate the initial solutions, and this method is very time-consuming in such large and complex grid maps.

On the contrary, we demonstrated that the relaxed version of A^* exhibits a substantial gain in terms of computational time but at the risk of missing optimal paths and obtaining longer paths. However, simulation results demonstrated that for the most of the tested problems, optimal or near-optimal path is reached (at most 10.1% of extra length and less than 0.4% on average).

We can conclude that RA^* algorithm is the best path planner as it provides a good trade-off of all metrics. Heuristic methods are not appropriate for grid path planning. Exact methods such as A^* cannot be used in large grid maps as they are

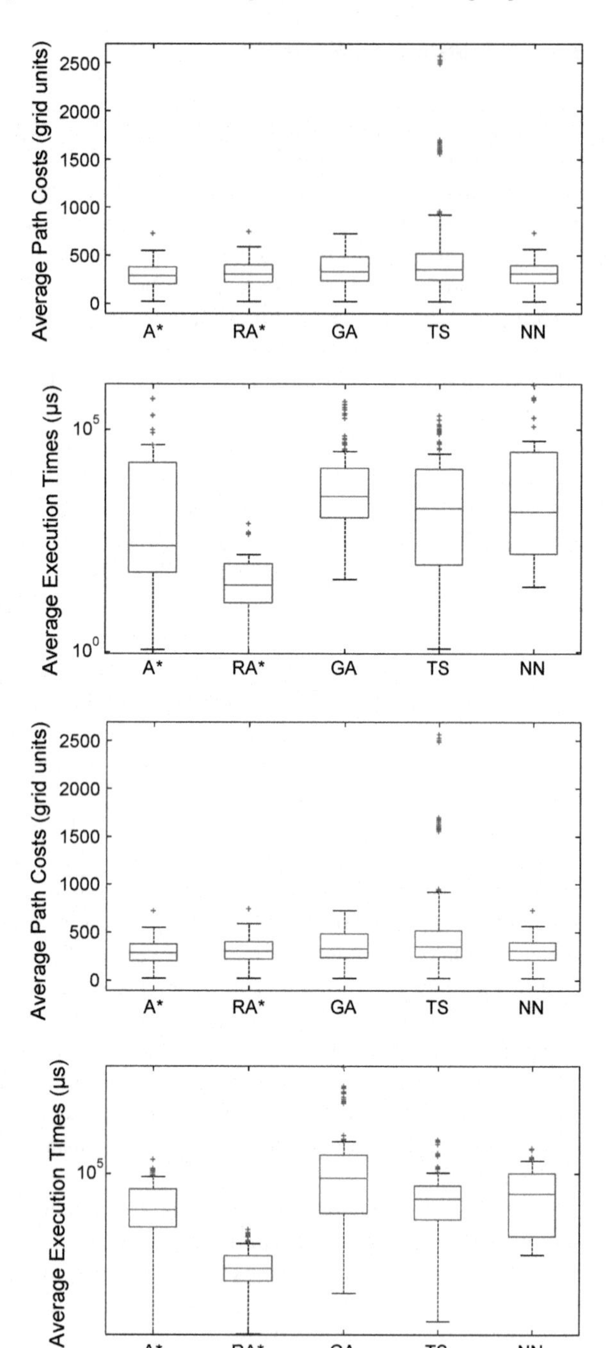

Fig. 3.6 Box plot of the average path costs and the average execution times (log scale) in 100 × 100, 500 × 500, and 1000 × 1000 random maps of heuristic approaches, Tabu Search, genetic algorithms, and neural network as compared to A* and RA*

Fig. 3.7 Box plot of the average path costs and the average execution times (log scale) in 512 × 512 random, 512 × 512 rooms, 512 × 512 video games, and 512 × 512 mazes maps of heuristic approaches Tabu Search, genetic algorithms, and neural network as compared to A* and RA*

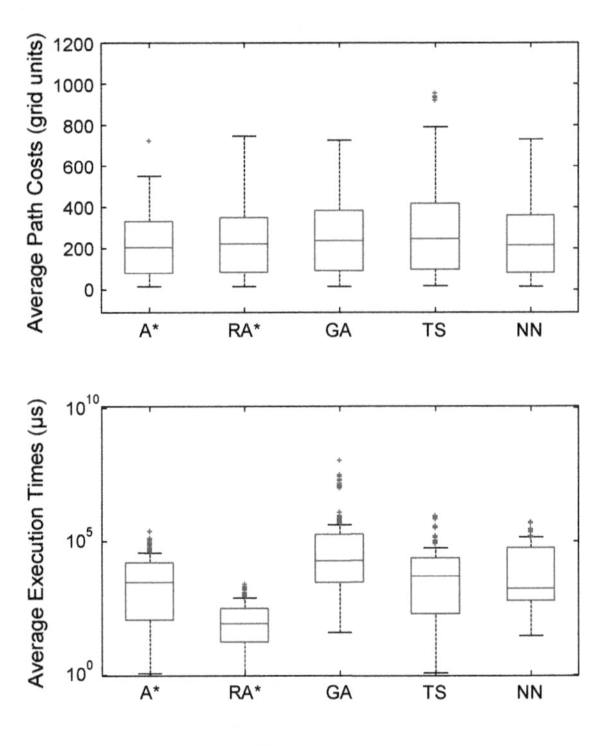

Fig. 3.8 Box plot of the average path costs and the average execution times (log scale) in the different maps (randomly generated and those of benchmark) of heuristic approaches Tabu Search, genetic algorithms, and neural network as compared to A* and RA*

Table 3.2 Average execution times (microseconds) for the different algorithms, per environment size

Algorithm	100 × 100	500 × 500	1000 × 1000	2000 × 2000	Random (512 × 512)	Rooms (512 × 512)	Video games (512 × 512)	Mazes (512 × 512)
A^*	114	9196	102647	296058499	17626	62753	11761	3161774
RA^*	17	67	271	15414	299	834	78	2435
GA	9687	34405	20329	–	2587005	849645	39462	–
TS	544	23270	39130	–	58605	144768	110478	–
NN	688	27713	321630	–	26211	62599	13802	–

time-consuming, and they can be used for small size problems or for short paths (near start and goal cells). However, heuristic methods have good features that can be used to improve the solution quality of near-optimal relaxed version of A^* without inducing extra execution time.

Lessons:

We have thoroughly analyzed and compared five path planners for mobile robot path planning, namely A^*, a relaxed version of A^*, the genetic algorithm, the Tabu Search algorithm, and the neural network algorithm. These algorithms pertain to two categories of path planning approaches: heuristic and exact methods. To demonstrate the feasibility of RA^* in real-world scenarios, we integrated it in the robot operating

Table 3.3 Percentage of extra length compared to optimal paths, calculated for non-optimal paths

Algorithm	100 × 100 (%)	500 × 500 (%)	1000 × 1000 (%)	2000 × 2000 (%)	Random (512 × 512) (%)	Rooms (512 × 512) (%)	Video games (512 × 512) (%)	Mazes (512 × 512) (%)
A^*	0.0	0.0	0.0	0.0	0.0	0.0	0.0	0.0
RA^*	6.99	0.4	1.97	6.81	5.48	10.13	5.95	2.356
GA	13.68	10.42	1.55	–	16.1	40.48	12.9	–
TS	35.74	17.7	13.06	–	47.31	51.75	26.82	–
NN	5.06	3.2	3.44	–	7.32	3.75	4.3	–

Table 3.4 Percentage of optimal paths, per environment size

Algorithm	100 × 100 (%)	500 × 500 (%)	1000 × 1000 (%)	2000 × 2000 (%)	Random (512 × 512) (%)	Rooms (512 × 512) (%)	Video games (512 × 512) (%)	Mazes (512 × 512) (%)
A^*	100	100	100	100	100	100	100	100
RA^*	5	40	60	10	5	0	20	55
GA	10	50	60	–	10	0	30	–
TS	5	0	0	–	5	0	0	–
NN	5	10	20	–	5	0	0	–

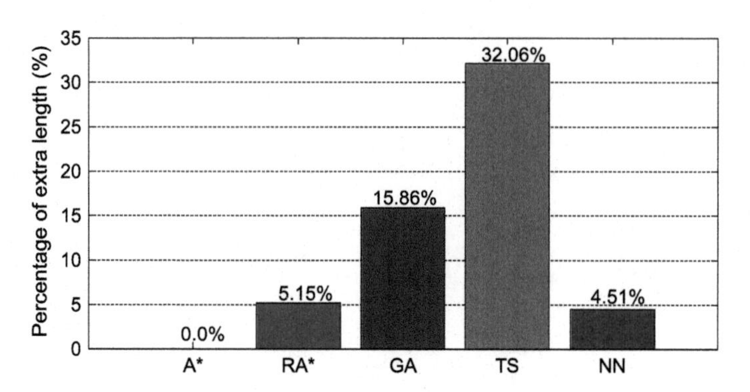

Fig. 3.9 Average Percentage of extra length compared to optimal path, calculated for non-optimal paths

system and we compared it against navfn in terms of path quality and execution time. We retain the following general lessons from the performance evaluation presented in the previous sections:

- **Lesson 1**: The study shows that heuristic algorithms are in general not appropriate for solving the path planning problem in grid environments. They are not as effective as A^* since the latter always exhibits the shortest execution times and the best solution qualities. GA was found to be less effective for large problem instances. It is able to find optimal solutions like A^* in some cases, but it always

exhibits a greater execution time. Tabu Search was also found to be the least effective as the exploration space is very huge in large problems, and it only explores the neighborhood of the initial solution. On the contrary, neural network provides better solutions than the two aforementioned techniques.
This can be explained by two reasons: The first reason is that heuristic approaches need to generate one or several initial paths with the greedy approach (in the case of GA and TS) and this operation itself takes time which is comparable to the execution of the whole A^* algorithm. The second reason is that heuristic approaches need several iterations to converge and this number of iteration depends on the complexity of the map and the positions of the start and goal cells.

- **Lesson 2**: The simulation study proved that exact methods in particular A^* have been found not appropriate for large grid maps. A^* requires a large computation time for searching path in such maps; for instance, in 2000 * 2000 grid map the execution time of A^* is around 3 hours, if we fix the start cell in the leftmost and topmost cell in the grid and the goal in the bottommost and rightmost cell in the grid. Thus, we can conclude that this type of path planning approaches can be used in real time only for small problem sizes (small grid maps) or for close start and goal cells.
- **Lesson 3**: Overall, RA^* algorithm is found to be the best path planner as it provides a good trade-off of all metrics. In fact, RA^* is reinforced by several mechanisms to quickly find good (optimal) solutions. For instance, RA^* exploits the grid structure and approximates the $g(n)$ function by the minimum-move path. Moreover, RA^* removes some unnecessary instructions used in A^* which contributes in radically reducing the execution time as compared to A^* without losing much in terms of path quality. It has been proved that RA^* can deal with large-scale path planning problems in grid environments in reasonable time and good results are obtained for each category of maps and for different couples of start and goal positions tested; in each case, a very near-optimal solution is reached (at most 10.1% of extra length and less than 0.4% in average) which makes it overall better than A^* and than heuristic methods.

The previous conclusions respond to the research question that we addressed in the iroboapp project about which method is more appropriate for solving the path planning problem. It seems from the results that heuristics methods including evolutionary computation techniques, such as GA, local search techniques, namely the Tabu Search, and neural networks cannot beat the A^* algorithm. A^* also is not appropriate to solve robot path planning problems in large grid maps as it is time-consuming, which is not convenient for robotic applications in which real-time aspect is needed. RA^* is the best algorithm in this study, and it outperforms A^* in terms of execution time at the cost of slightly losing optimal solution. Thus, we designed new hybrid approaches that take the benefits of both RA^* and heuristic approaches in order to ameliorate the path cost without inducing significant extra execution time.

3.5 Hybrid Algorithms for Robot Path Planning

As mentioned in the previous section, RA^* was found to be the most appropriate algorithm in this study among the different studied algorithms. Heuristic and exact methods are not suitable to solve global path planning problem for large grid maps. However, heuristic methods have good features that can be used to improve the near-optimal solutions of RA^*without inducing too much extra execution time. This observation led us to design new hybrid approaches by using RA^* to generate an initial path, which is further optimized by using a heuristic method, namely GA, Tabu Search, and ACO. In this section, we will demonstrate through simulations the validity of our intuition about the effectiveness of the hybrid techniques to simultaneously improve the solution quality and reduce the execution time.

3.5.1 *Design of Hybrid Path Planners*

The key idea of the hybrid approach consists of combining the RA^* and one heuristic method together. The hybrid algorithm comprises two phases: (i) the initialization of the algorithm using RA^* and (ii) a post-optimization phase (or local search) using one heuristic method that improves the quality of solution found in the previous phase. We designed three different hybrid methods: $RA^* + TS$, $RA^* + ACO$, and $RA^* + GA$ aiming at comparing their performances and choosing the appropriate one.

3.5.1.1 Initialization Using RA^*

As it is described in the previous section, the use of the greedy approach to generate the initial path(s) in the case of GA and TS increases the execution time of the whole algorithm, especially in large grid maps. This led us to use RA^* instead of the greedy approach in the quest of ensuring a fast convergence toward the best solution.

The hybrid $RA^* + TS$ algorithm will consider the RA^* path as an initial solution.

In the original ACO algorithm, an initial pheromone value is affected to the transition between the cells of the grid map. After each iteration of the algorithm, the quantities of pheromone are updated by all the ants that have built paths. The pheromone values increase on the best paths during the search process. The core idea of RA* + ACO is to increase the quantities of pheromone around the RA^* path from the beginning of the algorithm in order to guide the ants toward the best path and to accelerate the search process. Thus, we consider different values of pheromones; the quantities of pheromone between the cells of the RA^* path and on their neighborhood at radius N (N is randomly generated) will be higher than the remaining cells in the maps.

Algorithm 10. The Hybrid Algorithm $RA^* + GA$

input : $Grid, Start, Goal$
1 **repeat**
2   **if** *size == 1* **then**
3   Generate the RA* path (described in Algorithm. 1)
4   Add the RA* to the current population.
5   **else**
6   Choose Randomly an intermediate cell not in RA*.
7     $Neighbors_{RadiusN}$: set of neighbors at Radius N, N is a parameter.
8     $Cross_{Point}$: Select randomly one cell from $N_{RadiusN}$.
9     $Path_{S,CP}$: find the path using RA* between start cell and $Cross_{Point}$.
10     $Path_{CP,G}$: find the path using RA* between $Cross_{Point}$ and the goal cell.
11   Add $Path_{S,CP} \cup Path_{CP,G}$ to the i population.
12   **end**
13 **until** *(size $\geq$ max population size)*;
14 **while** *(generation number < max generation number)* **do**
15   Use GA to generate the next population.
16 **end**
17 Generate the best path.

In $RA^* + GA$, RA^* algorithm will be used to generate the initial population of the GA algorithm. The generation of the initial population starts with generating an initial path from the start cell to the goal cell using the RA^* algorithm. To generate the subsequent paths in the initial population, the algorithm will choose a random intermediate cell, not in the RA^* path, which will be used to generate a new path from start to goal positions across the selected intermediate cell.

3.5.1.2 Post-Optimization Using Heuristic Methods

This phase is a kind of post-optimization or local search. It consists of improving the quality of solution found in the first phase using one heuristic method among GA, TS, and ACO. We used different heuristic methods in order to compare between their performance. Because of the limited space for this chapter, we are not able to present all the hybrid algorithms and only the $RA^* + GA$ algorithm is presented in Algorithm 11. The flowchart diagram is depicted in Fig. 3.10.

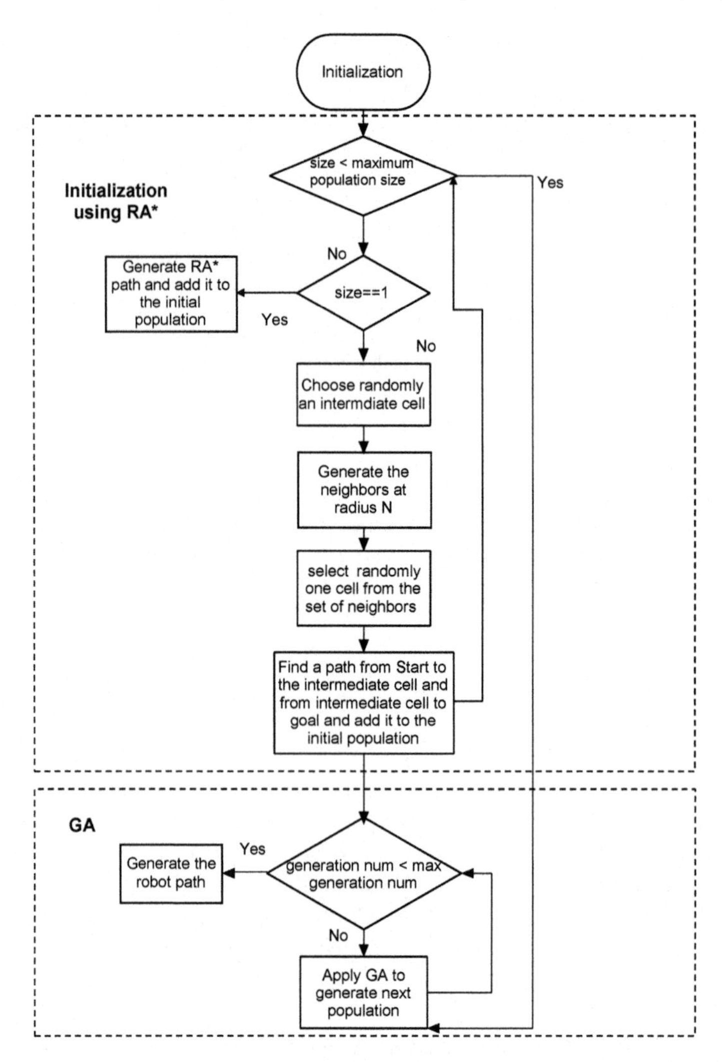

Fig. 3.10 Flowchart diagram of the RA* + GA hybrid algorithm

3.5.2 Performance Evaluation

In this section, we present the simulation results of the hybrid algorithms. To evaluate the performance of the hybrid algorithms, we compared them against A^* and RA^*. Two performance metrics were assessed: the path length and the execution time of each algorithm. Figure 3.11, Tables 3.5 and 3.6 present the average path costs and the average execution times of A^*, RA^*, $RA^* + GA$, and $RA^* + TS$ algorithms in different kinds of maps. Looking at these figures, we notice that A^*always exhibits the shortest path cost and the longest execution time. Moreover, we clearly observe that

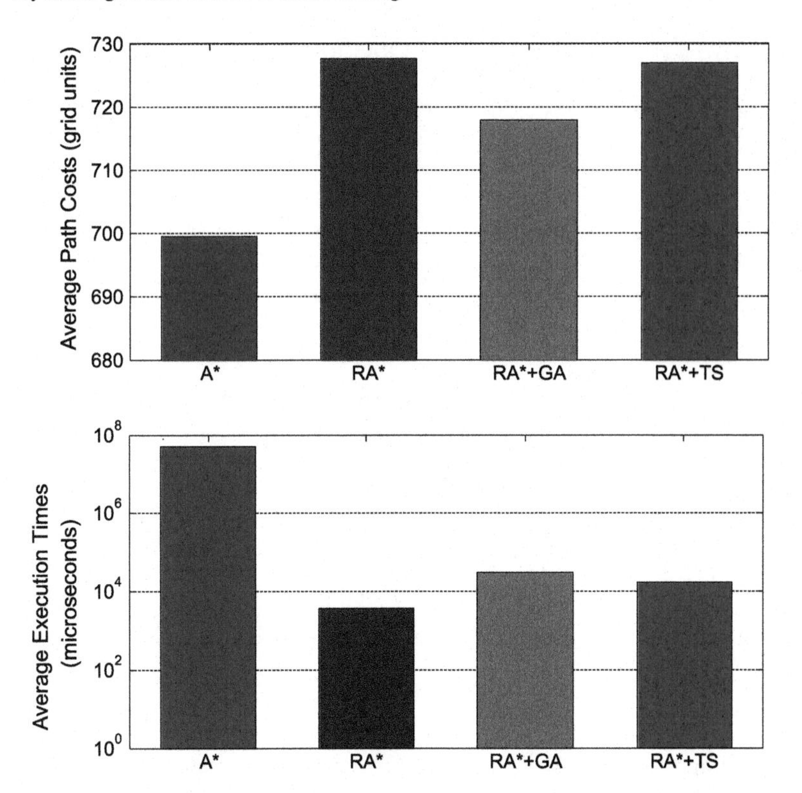

Fig. 3.11 Average path lengths and average execution times (log scale) of hybrid approach RA*+GA and RA* + TS as compared to A* and RA*

RA^*+GA hybridization provides a good trade-off between the execution time and the path quality. In fact, the RA^*+GA hybrid approach provides better results than RA^* and reduces the execution time as compared to A^* for the different types of maps. This confirms the benefits of using hybrid approaches for post-optimization purposes for large-scale environments. However, the hybrid approach using RA^* and Tabu Search provides some improvements but less significant than the $RA^* + GA$ approach. In some cases, $RA * +TS$ could not improve the RA^* path cost (in video games maps and 2000×2000 grid maps). This can be explained by the fact that Tabu Search approach only explores the neighborhood of the RA^* solution initially generated by applying simple moves (remove, exchange, insert) which cannot significantly improve it. We can see also that all the non-optimal solutions have a small gap. Finally, the hybrid approach $RA^* + ACO$ was not successful as it does not converge easily to a better solution that the initial RA^* algorithm because of the randomness nature of the ant motions and its execution time remains not interesting.

We designed a new hybrid algorithm that combines both RA^* and GA. The first phase of the $RA^* + GA$ consists to use RA^* to generate the initial population of GA instead of the greedy approach in order to reduce the execution time, and then, GA

Table 3.5 Average path costs (grid units) for A*, RA*, RA* + GA, and RA* + TS algorithms, per environment size

Algorithm	100 × 100	2000 × 2000	Random (512 × 512)	Rooms (512 × 512)	Video games (512 × 512)	Mazes (512 × 512)
A^*	66.2	1366.4	303.4	319.1	479.5	1661.9
RA^*	70.5	1443.3	319.5	350.9	493.8	1687.1
RA* + GA	68.2	1423.4	315.3	335.0	486.1	1679.2
RA* + TS	69.3	1443.3	319.5	348.2	493.8	1687.1

Table 3.6 Average execution times (microseconds) for A*, RA*, RA* + GA, and RA* + TS, per environment size

Algorithm	100 × 100	2000 × 2000	Random (512 × 512)	Rooms (512 × 512)	Video games (512 × 512)	Mazes (512 × 512)
A^*	114	296058499	17626	73438	38649	3161774
RA^*	17	15414	299	724	436	5220
RA* + GA	69	59149	1295	1813	7160	116807
RA* + TS	110	22258	4298	8900	16174	51375

is used to improve the path quality found in the previous phase. We demonstrated through simulation that the hybridization between RA^* and GA brings a lot of benefits as it gathers the best features of both approaches, which contributes to improve the solution quality as compared to RA^* and reduce the search time for large-scale graph environments as compared to A^*.

3.6 Conclusion

To answer to the research question that we addressed in the iroboapp project, we designed and compared in this chapter five different path planners used to solve global path planning problem in large grid maps. The algorithms considered are A^*, a relaxed version of A^* called RA^*, the genetic algorithm, the Tabu Search algorithm, and the neural network algorithm. The obtained results show that RA^* is the best algorithm as it provides a good trade-off between solution quality and execution time. In the next chapter, we will demonstrate the feasibility and effectiveness of RA^* planner in real-world scenarios by integrating it as global path planner in the Robot Operating System (ROS) [18] as possible replacement of the default navfn path planner (based on a variant of the Dijkstra algorithm).

References

1. Anis Koubaa. 2014. The Iroboapp Project. http://www.iroboapp.org. Accessed 27 Jan 2016.
2. Morgan Quigley, Ken Conley, Brian Gerkey, Josh Faust, Tully Foote, Jeremy Leibs, Rob Wheeler, and Andrew Y Ng. Ros. 2009. An open-source robot operating system. In *ICRA workshop on open source software*, vol. 3, p. 5. Kobe.
3. Anis Koubaa. 2014. Ipath simulator. http://www.iroboapp.org/index.php?title=IPath. Accessed 6 Nov 2014.
4. Anis Koubaa. 2014. Api documentation. http://www.iroboapp.org/ipath/api/docs/annotated.html. Accessed 6 Nov 2014.
5. Evangelos Kanoulas, Yang Du, Tian Xia, and Donghui Zhang. 2006. Finding fastest paths on a road network with speed patterns. In *Proceedings of the 22nd international conference on data engineering, ICDE 2006*, 10–10. IEEE.
6. Ammar, Adel, Hachemi Bennaceur, Imen Châari, Anis Koubâa, and Maram Alajlan. 2016. Relaxed dijkstra and a* with linear complexity for robot path planning problems in large-scale grid environments. *Soft Computing* 20 (10): 4149–4171.
7. Russell, S., and P. Norvig. 2009. *Artificial intelligence: a modern approach*, 3rd ed. Prentice Hall.
8. Huilai Zou, Lili Zong, Hua Liu, Chaonan Wang, Zening Qu, and Youtian Qu. 2010. Optimized application and practice of a* algorithm in game map path-finding. In *IEEE 10th international conference on computer and information technology (CIT)*, 2138–2142. IEEE.
9. Imen Châari, Anis Koubâa, Hachemi Bennaceur, Adel Ammar, Sahar Trigui, Mohamed Tounsi, Elhadi Shakshuki, and Habib Youssef. 2014. On the adequacy of tabu search for global robot path planning problem in grid environments. *Procedia Computer Science*, 32(0): 604–613, *The 5th international conference on ambient systems, networks and technologies (ANT-2014), the 4th international conference on sustainable energy information technology (SEIT-2014).*
10. Maram Alajlan, Anis Koubaa, Imen Chaari, Hachemi Bennaceur, and Adel Ammar. 2013. Global path planning for mobile robots in large-scale grid environments using genetic algorithms. In *2013 international conference on individual and collective behaviors in robotics ICBR'2013*, Sousse, Tunisia.
11. Alajlan, Maram, Imen Chaari, Anis Koubaa, Hachemi Bennaceur, Adel Ammar, and Habib Youssef. 2016. Global robot path planning using ga for large grid maps: modelling, performance and experimentation. *International Journal of Robotics and Automation* 31: 1–22.
12. Adem Tuncer and Mehmet Yildirim. 2011. Chromosome coding methods in genetic algorithm for path planning of mobile robots. In *Computer and Information Sciences II*, 377–383. Springer.
13. Masoud Samadi and Mohd Fauzi Othman. 2013. Global path planning for autonomous mobile robot using genetic algorithm. In *International conference on signal-image technology & internet-based systems (SITIS)*, 726–730. IEEE.
14. Imen Châari, Anis Koubaa, Hachemi Bennaceur, Sahar Trigui, and Khaled Al-Shalfan. 2012. Smartpath: A hybrid aco-ga algorithm for robot path planning. In *IEEE congress on evolutionary computation (CEC)*, 1–8. IEEE.
15. Mandavilli Srinivas and Lalit. M Patnaik. 1994. *Genetic algorithms: A survey. computer* 27 (6): 17–26.
16. Nathan Sturtevant. 2012. Benchmark. http://www.movingai.com/benchmarks/.
17. Anis Koubaa. 2014. Grid-maps: 10 x 10 up to 2000 x 2000. http://www.iroboapp.org/index.php?title=Maps. Accessed 28 Jan 2014.
18. Robot Operating System (ROS). http://www.ros.org.

Chapter 4
Integration of Global Path Planners in ROS

Abstract Global path planning consists in finding a path between two locations in a global map. It is a crucial component for any map-based robot navigation. The navigation stack of the Robot Operating System (ROS) open-source middleware incorporates both global and local path planners to support ROS-enabled robot navigation. Only two basic algorithms are defined for the global path planner including Dijkstra and carrot planners. However, more intelligent global planners have been defined in the literature but were not integrated in ROS distributions. The contribution of this work consists in integrating the RA^* path planner, defined in Chap. 3, into the ROS global path planning component as a plugin. We demonstrate how to integrate new planner into ROS and present their benefits. Extensive experimentations are performed on simulated and real robots to show the effectiveness of the newly integrated planner as compared to ROS default planner.

4.1 Introduction

In this section, we present a general overview of the basic concepts of the Robot Operating System (ROS) framework [1]. This overview is not intended to be comprehensive but just an introduction of the important concepts, and interested readers may refer to [2] for details.

Robot Operating System (ROS) has been developed, by Willow Garage [3] and Stanford University as a part of STAIR [4] project, as a free and open-source robotic middleware for the large-scale development of complex robotic systems.

ROS acts as a metaoperating system for robots as it provides hardware abstraction, low-level device control, inter-processes message-passing and package management. It also provides tools and libraries for obtaining, building, writing, and running code across multiple computers. The main advantage of ROS is that it allows manipulating sensor data of the robot as a labeled abstract data stream, called topic, without having to deal with hardware drivers. This makes the programming of robots much easier for software developers as they do not have to deal with hardware drivers and interfaces. Also, ROS provides many high-level applications such as arm controllers, face tracking, mapping, localization, and path planning. This allows the researchers

A. Koubaa et al., *Robot Path Planning and Cooperation*, Studies in Computational Intelligence 772, https://doi.org/10.1007/978-3-319-77042-0_4

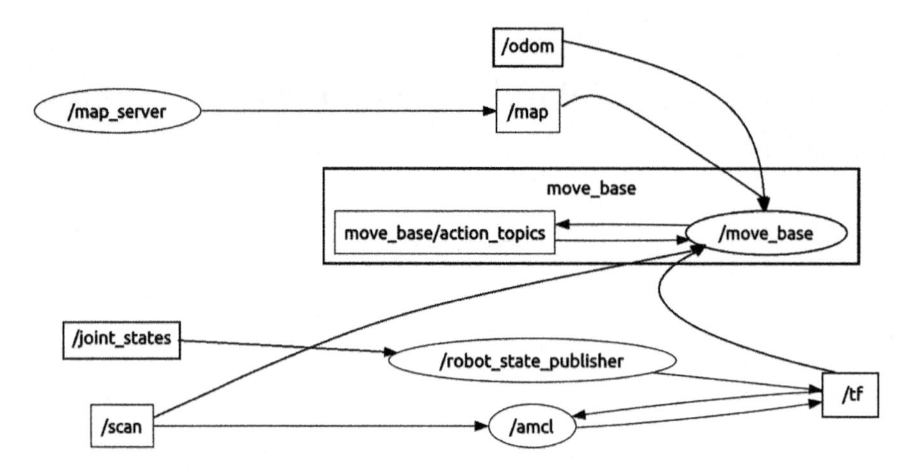

Fig. 4.1 Example of a ROS computation graph

to focus on specific research problems rather than on implementing the many necessary, but unrelated parts of the system. It is useful to mention that ROS is not a real-time framework, though it is possible to integrate it with real-time code.

ROS relies on the concept of computational graph, which represents the network of ROS processes (potentially distributed across machines). An example of a simplified computation graph is illustrated in Fig. 4.1.

A process in ROS is called a *node*, which is responsible for performing computations and processing data collected from sensors. As illustrated in Fig. 4.1, a ROS system is typically composed of several nodes (i.e., processes), where each node processes a certain data. For example, *move_base* is a node that controls the robot navigation, *amcl* is another node responsible for the localization of the robot, and *map_server* is a node that provides the map of the environment to other processes of the system. Nodes are able to communicate through message passing, where a *message* is a data structure with different typed fields. This communication between nodes is only possible thanks to a central node, referred to as *ROS Master*, which acts as a name server providing name registration and lookup for all components of a ROS computation graph (e.g., nodes), and stores relevant data about the running system in a central repository called *Parameter Server*.

ROS supports two main communication models between nodes:

- The *publish/subscribe* model: In this model, nodes exchange *topics*, which represents a particular flow on data. One node or several nodes may act as a publisher(s) of a particular topic, and several nodes may subscribe to that topic, through the *ROS Master*. Subscriber and publisher nodes do not need to know about the existence between other because the interaction is based on the topic name and made through the *ROS Master*. For example, in Fig. 4.1, the *map_server* is the publisher of the topic */map*, which is consumed by the subscriber node *move_base*, which uses the map for navigation purposes. */scan* represents the

flow of data received from the laser range finder, also used by *move_base* node to avoid obstacles. */odom* represents the control information used by *move_base* to control robot motion.
- The *request/reply* model: In this model, one node acts as a server that offers the service under a certain name and receives and processes requests from other nodes acting as clients. Services are defined by a pair of message structures: one message for the request, and one message for the reply. Services are not represented in the ROS computation graph.

Mobile robot navigation generally requires solutions for three different problems: mapping, localization, and path planning. In ROS, the *Navigation Stack* plays such a role to integrate together all the functions necessary for autonomous navigation. In the next section, we will present an overview about the navigation stack.

4.2 Navigation Stack

In order to achieve the navigation task, the *Navigation Stack* [5] is used to integrate the mapping, localization, and path planning together. It takes information from odometry, sensor streams, and the goal position to produce safe velocity commands and send it to the mobile base [1]. The odometry comes through `nav_msgs/Odometry` message over ROS which stores an estimate of the position and velocity of a robot in free space to determine the robot's location. The sensor information comes through either `sensor_msgs/LaserScan` or `sensor_msgs/PointCloud` messages over ROS to avoid any obstacles. The goal is sent to the navigation stack by `geometry_msgs/PoseStamped` message. The navigation stack sends the velocity commands through `geometry_msgs/Twist` message on `/cmd_vel` topic. The `Twist` message composed of two submessages:

```
1  geometry_msgs/Vector3 linear
2      float64 x
3      float64 y
4      float64 z
5  geometry_msgs/Vector3 angular
6      float64 x
7      float64 y
8      float64 z
```

`linear` submessage is used for the x, y, and z linear velocity components in meters per second and `angular` submessage is used for the x, y, and z angular velocity components in radians per second. For example, the following `Twist` message:

```
linear: {x: 0.2, y: 0, z: 0}, angular: {x: 0, y: 0, z: 0}
```

will tell the robot to move with a speed of 0.2 m/s straight ahead. The base controller is responsible for converting `Twist` messages to "motor signals" which will actually move the robot's wheels [6].

The navigation stack does not require a prior static map to start with. Actually, it could be initialized with or without a map. When initialized without a prior map, the robot will know about the obstacles detected by its sensors only and will be able to avoid the seen obstacles so far. For the unknown areas, the robot will generate an optimistic global path which may hit unseen obstacles. The robot will be able to re-plan its path when it receives more information by the sensors about these unknown areas. Instead, when the navigation stack initialized with a static map for the environment, the robot will be able to generate an informed plans to its goal using the map as prior obstacle information. Starting with a prior map will have significant benefits on the performance [5].

To build a map using ROS, ROS provides a wrapper for OpenSlam's Gmapping [7]. A particle filter-based mapping approach [8] is used by the `gmapping` package to build an occupancy grid map. Then, a package named `map_server` could be used to save that map. The maps are stored in a pair of files: `YAML` file and image file. The `YAML` file describes the map metadata and names the image file. The image file encodes the occupancy data. The localization part is solved in the `amcl` package using an Adaptive Monte Carlo Localization [9] which is also based on particle filters. It is used to track the position of a robot against a known map. The path planning part is performed in the `move_base` package and is divided into *global* and *local* planning modules which is a common strategy to deal with the complex planning problem.

The *global path planner* searches for a shortest path to the goal and the *local path planner* (also called the *controller*), incorporating current sensor readings, issues the actual commands to follow the global path while avoiding obstacles. More details about the global and local planners in ROS can be found in the next sections.

The *move_base* package also maintains two costmaps, *global_costmap* and *local_costmap*, to be used with the global and local planners, respectively. The costmap used to store and maintain information in the form of occupancy grid about the obstacles in the environment and where the robot should navigate. The costmap initialized with prior static map if available, then it will be updated using sensor data to maintain the information about obstacles in the environment. Besides that, the *move_base* may optionally perform some previously defined recovery behaviors (Fig. 4.2) when it fails to find a valid plan.

One reason of failure is when the robot find itself surrounded with obstacles and cannot find a way to its goal. The recovery behaviors will be performed in some order (defined by the user), and after performing one recovery, the *move_base* will try to find a valid plan, if it succeeds, it will proceed its normal operation. Otherwise if it fails, it will perform the next recovery behavior. If it fails after performing all the recovery behaviors, the goal will be considered infeasible, and it will be aborted. The default recovery behaviors order is presented in Fig. 4.2 [1] and it is in increasingly aggressive order to attempt to clear out the robot space. First recovery behavior is clearing all the obstacles outside a specific area from the robot's map. Next, an

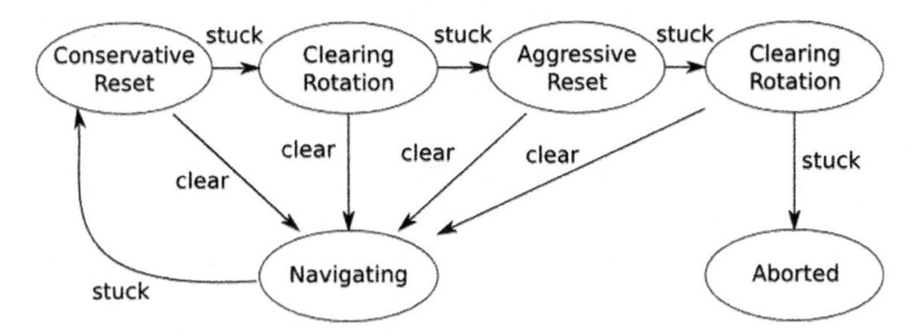

Fig. 4.2 Recovery behaviors

in-place rotation will be performed if possible to clear the space. More aggressively clearing for the map will be performed, in case of fail of the two recovery behaviors mentioned, to remove all the obstacles outside of the rectangular area in which the robot can make an in-place rotation. Next, another in-place rotation will be performed. If all this fails, the goal will be aborted.

Therefore, in each execution cycle of the *move_base*, one of three main states should be performed:

- Planning state: run the *global path planner*.
- Controlling state: run the *local path planner* and move the robot.
- Clearing state: run recovery behavior in case the robot stuck.

There are some predefined parameters in ROS navigation stack that are used to control the execution of the states, which are:

- *planner_frequency*: to determine how often the *global path planner* should be called and is expressed in *Hz*. When it is set to zero, the global plan will be computed only once for each goal received.
- *controller_frequency*: to determine how often the *local path planner* or *controller* should be called and also expressed in *Hz*.

For any global or local planner or recovery behavior to be used with the *move_base*, it must be first adhere to some interfaces defined in *nav_core* package, which contains key interfaces for the navigation stack, then it must be added as a plugin to ROS. We developed a tutorial on how to add a new global planner as a plugin to ROS navigation stack, available at [10, 11].

4.2.1 Global Planner

The *global path planner* in ROS operates on the *global_costmap*, which generally initialized from a prior static map, then it could be updated frequently based on the value of *update_frequency* parameter. The *global path planner* is responsible for generating a long-term plan from the start or current position to the goal position before the robot starts moving. It will be seeded with the costmap, and the start and goal positions. These start and goal positions are expressed by their x and y coordinates. A grid-based global planner that can use Dijkstra's algorithm [12] or A* algorithm [13] to compute shortest collision-free path for a robot is obtained in `global_planner` package. Also, ROS provides another global planner named `carrot_planner`, which is a simple planner that attempts to move the robot as close to its goal as possible even when that goal is in an obstacle. The current implementation of the global planner in ROS assumes a circular-shape robot. This results in generating an optimistic path for the actual robot footprint, which may be infeasible path. Besides that, the global planner ignores kinematic and acceleration constraints of the robot, so the generated path could be dynamically infeasible.

4.2.2 Local Planner

The *local path planner* or the *controller* in ROS operates on the *local_costmap*, which only uses local sensor information to build an obstacle map and dynamically updated with sensor data. It takes the generated plan from the global planner, and it will try to follow it as close as possible considering the kinematics and dynamics of the robot as well as any moving obstacles information in the *local_costmap*. ROS provides implementation of two local path planning algorithms namely the Trajectory Rollout [14] and the Dynamic Window Approach (DWA) [15] in the package *base_local_planner*. Both algorithms have the same idea to first discretely sampled the control space then to perform forward simulation, and the selection among potential commands. The two algorithms differ in how they sample the robot's control space.

After the global plan passed to the *controller*, the *controller* will produce velocity commands to send to a mobile base. For each control cycle, *the controller* will try to process a part from global path (determined by the size of the *local_costmap*).

First, the *controller* will sample the control space of the robot discretely. The number of the samples will be specified by the controller parameters *vx_samples* and *vtheta_samples* (more details about the parameters can be found in the next section). Then, the controller will perform a simulation in advance for each one of those velocity samples from the current place of the robot to foresee the situation from applying each sample for amount of time (this time will be specified in the parameter *sim_time*). Then, the controller will evaluate each resultant path from the simulation and will exclude any path having collisions with obstacles. For the evaluation, the controller will incorporate the following metrics: distance from obstacles, distance to the goal position, distance from the global plan and robot speed. Finally, the controller will send the velocity command of the highest-scoring path to the mobile base to execute it.

The "Map Grid" is used to evaluate and score the velocities. For each control cycle, the controller will create a grid around the robot (the grid size determined by the size of the *local_costmap*), and the global path will be mapped onto this area. Then, each grid cell will receive a distance value. The cells containing path points and the goal will be marked with 0. Then, each other grid cell will be marked with its manhattan distance from nearest zero grid by a propagation algorithm. This "Map Grid" is then used in the evaluation and scoring of the velocities. As the "Map Grid" will cover small area from global path each time, the goal position often will lie outside that area. So in that case the first path point inside the area having a consecutive point outside the area will be considered as "local goal," and the distance from that local goal will be considered when scoring trajectories for distance to goal.

4.3 How to Integrate a New Path Planner as Plugin?

Once the global path planner class is written, it must be added as a plugin to ROS so that it can be used by the *move_base* package. First, the global planner class has to be registered as plugin by exporting it. The class needs to be labeled as an exported class to allow it to be dynamically loaded. This can be done using the macro *PLUGINLIB_EXPORT_CLASS*, which usually placed at the end of the source file (*.cpp*) of the exported class. Note that, it can be placed at any source file (*.cpp*) from the plugin library. The second step is to write the plugin description file, which is an *XML* file and used to keep all the information about the plugin such as the plugin name, type. This plugin description file must be pointed in the export tag block inside the *package.xml*. Each package should specify what are the plugins it exports besides to the package libraries contain those plugins. This is needed to allow *pluginlib* to query the available plugins across all ROS packages.

In what follows, we present the steps of integrating a new path planner into ROS. The integration has two main steps: (1) writing the path planner class and (2) deploying it as a plugin. Following, we describe them in details.

4.3.1 *Writing the Path Planner Class*

As mentioned before, to make a new global planner work with ROS, it must first adhere to the interfaces defined in `nav_core` package. Thus, all the methods in the class `nav_core::BaseGlobalPlanner` must be overridden. This section presents how to write a new global path planner consisting of two file: (1) header file `RAstar_ros.h`, and (2) source file `RAstar_ros.cpp`. Following is the header content:

```
1  /** include the libraries you need in your planner here */
2  /** for global path planner interface */
3  #include <ros/ros.h>
4  #include <costmap_2d/costmap_2d.h>
5  #include <costmap_2d/costmap_2d_ros.h>
6  #include <nav_core/base_global_planner.h>
```

```
7  #include <geometry_msgs/PoseStamped.h>
8  #include <angles/angles.h>
9  #include <base_local_planner/world_model.h>
10 #include <base_local_planner/costmap_model.h>
11
12 using std::string;
13
14 #ifndef RASTAR_ROS_CPP
15 #define RASTAR_ROS_CPP
16
17 namespace RAstar_planner {
18
19 class RAstarPlannerROS : public nav_core::BaseGlobalPlanner {
20 public:
21
22  RAstarPlannerROS();
23  RAstarPlannerROS(std::string name, costmap_2d::Costmap2DROS* costmap_ros);
24
25  /** overridden classes from interface nav_core::BaseGlobalPlanner **/
26  void initialize(std::string name, costmap_2d::Costmap2DROS* costmap_ros);
27  bool makePlan(const geometry_msgs::PoseStamped& start,
28          const geometry_msgs::PoseStamped& goal,
29          std::vector<geometry_msgs::PoseStamped>& plan
30             );
31  };
32 };
33 #endif
```

First we have to include core ROS libraries needed for path planner. In line 4 and 5, we include the headers needed for `costmap_2d::Costmap2D` class which are `costmap_2d/costmap_2d.h` and `costmap_2d/costmap_2d_ros.h`. The `costmap_2d::Costmap2D` class will be used as input map by the path planner. Note that when defined the path planner class as a plugin, it will access this map automatically. Thus, no subscription to `costmap2d` is needed to get the cost map from ROS. Line 6 imports the interface `nav_core::BaseGlobalPlanner`, which any plugin must adhere to.

```
17 namespace RAstar_planner {
18
19 class RAstarPlannerROS : public nav_core::BaseGlobalPlanner {
```

The namespace `RAstar_planner` is defined for the class `RAstarPlanner ROS` in line 17. With the namespace, a full reference to the class can be defined such as `RAstar_planner::RAstarPlannerROS`. The class `RAstarPlannerROS` which inherits from the interface `nav_core::BaseGlobalPlanner` as mentioned before is defined in line 19. The new class `RAstarPlannerROS` must override all the methods defined in `nav_core::BaseGlobalPlanner`. The default constructor is in line 22 which used to initialize the attributes of planner with the default values. The constructor in line 23 is used to initialize the name of the planner and the costmap. Line 26 defined an initialization function for the `BaseGlobalPlanner`, which used to initialize the map to be used for the planning and planner name. The initialize method for RA* is implemented as follows:

```
void RAstarPlannerROS::initialize(std::string name, costmap_2d::Costmap2DROS* costmap_ros)
{
    if (!initialized_)
    {
        costmap_ros_ = costmap_ros;
        costmap_ = costmap_ros_->getCostmap();
        ros::NodeHandle private_nh("~/" + name);

        originX = costmap_->getOriginX();
        originY = costmap_->getOriginY();
        width = costmap_->getSizeInCellsX();
        height = costmap_->getSizeInCellsY();
        resolution = costmap_->getResolution();
        mapSize = width*height;
        tBreak = 1+1/(mapSize);
        OGM = new bool [mapSize];

        for (unsigned int iy = 0; iy < height; iy++)
        {
            for (unsigned int ix = 0; ix < width; ix++)
            {
                unsigned int cost = static_cast<int>(costmap_->getCost(ix, iy));
                if (cost == 0)
                    OGM[iy*width+ix]=true;
                else
                    OGM[iy*width+ix]=false;
            }
        }
        ROS_INFO("RAstar planner initialized successfully");
        initialized_ = true;
    }
    else
        ROS_WARN("This planner has already been initialized... doing nothing");
}
```

```
27 bool makePlan(const geometry_msgs::PoseStamped& start,
28         const geometry_msgs::PoseStamped& goal,
29         std::vector<geometry_msgs::PoseStamped>& plan
30            );
```

Then, the method bool `makePlan` must be overridden. The final plan will be stored in the parameter `std::vector<geometry_msgs::PoseStamped>& plan` of the method. This plan will be automatically published through the plugin as a topic.

4.3.1.1 Class Implementation

Here, we present the main issues in implementing a global planner as plugin. Following is part of the code from RA* global path planner[1]:

```
1 #include <pluginlib/class_list_macros.h>
2 #include "RAstar_ros.h"
3
```

[1]The complete source code is available in [16].

```
4  //register this planner as a BaseGlobalPlanner plugin
5  PLUGINLIB_EXPORT_CLASS(RAstar_planner::RAstarPlannerROS, nav_core::BaseGlobalPlanner)
6
7  using namespace std;
8
9  namespace RAstar_planner {
10 //Default Constructor
11 RAstarPlannerROS::RAstarPlannerROS(){
12
13 }
14 RAstarPlannerROS::RAstarPlannerROS(std::string name, costmap_2d::Costmap2DROS* costmap_ros){
15   initialize(name, costmap_ros);
16 }
17 void RAstarPlannerROS::initialize(std::string name, costmap_2d::Costmap2DROS* costmap_ros){
18
19 }
20 bool RAstarPlannerROS::makePlan(const geometry_msgs::PoseStamped& start, const geometry_msgs↩
      ::PoseStamped& goal,
21     std::vector<geometry_msgs::PoseStamped>& plan ){
22 if (!initialized_) {
23   ROS_ERROR("The planner has not been initialized, please call initialize() to use the ↩
         planner");
24   return false;
25  }
26  ROS_DEBUG("Got a start: %.2f, %.2f, and a goal: %.2f, %.2f", start.pose.position.x, start.↩
       pose.position.y,
27          goal.pose.position.x, goal.pose.position.y);
28  plan.clear();
29  if (goal.header.frame_id != costmap_ros_->getGlobalFrameID()){
30    ROS_ERROR("This planner as configured will only accept goals in the %s frame, but a goal ↩
         was sent in the %s frame.",
31            costmap_ros_->getGlobalFrameID().c_str(), goal.header.frame_id.c_str());
32    return false;
33  }
34  tf::Stamped < tf::Pose > goal_tf;
35  tf::Stamped < tf::Pose > start_tf;
36
37  poseStampedMsgToTF(goal, goal_tf);
38  poseStampedMsgToTF(start, start_tf);
39
40  // convert the start and goal coordinates into cells indices to be used with RA* planner
41  float startX = start.pose.position.x;
42  float startY = start.pose.position.y;
43  float goalX = goal.pose.position.x;
44  float goalY = goal.pose.position.y;
45
46  getCorrdinate(startX, startY);
47  getCorrdinate(goalX, goalY);
48
49  int startCell;
50  int goalCell;
51
52  if (isCellInsideMap(startX, startY) && isCellInsideMap(goalX, goalY)){
53    startCell = convertToCellIndex(startX, startY);
54    goalCell = convertToCellIndex(goalX, goalY);
55  }
56  else {
57    cout << endl << "the start or goal is out of the map" << endl;
58    return false;
59  }
60  /////////////////////////////////////////////////////////
61  // call RA* path planner
62  if (GPP->isStartAndGoalCellsValid(OGM, startCell, goalCell)){
63    vector<int> bestPath;
64    bestPath = RAstarPlanner(startCell, goalCell); // call RA*
65
66    //if the global planner find a path
```

```
67      if ( bestPath->getPath().size()>0)
68      {
69       // convert the path cells indices into coordinates to be sent to the move base
70        for (int i = 0; i < bestPath->getPath().size(); i++){
71          float x = 0.0;
72          float y = 0.0;
73          int index = bestPath->getPath()[i];
74
75          convertToCoordinate(index, x, y);
76
77          geometry_msgs::PoseStamped pose = goal;
78          pose.pose.position.x = x;
79          pose.pose.position.y = y;
80          pose.pose.position.z = 0.0;
81          pose.pose.orientation.x = 0.0;
82          pose.pose.orientation.y = 0.0;
83          pose.pose.orientation.z = 0.0;
84          pose.pose.orientation.w = 1.0;
85
86          plan.push_back(pose);
87        }
88        // calculate path length
89        float path_length = 0.0;
90        std::vector<geometry_msgs::PoseStamped>::iterator it = plan.begin();
91        geometry_msgs::PoseStamped last_pose;
92        last_pose = *it;
93        it++;
94        for (; it!=plan.end(); ++it) {
95            path_length += hypot( (*it).pose.position.x - last_pose.pose.position.x, (*it).pose. ↩
                  position.y - last_pose.pose.position.y );
96            last_pose = *it;
97        }
98        cout <<"The global path length: "<< path_length<< " meters"<<endl;
99        return true;
100     }
101     else{
102       cout << endl << "The planner failed to find a path " << endl
103           << "Please choose other goal position, " << endl;
104       return false;
105     }
106   }
107   else{
108     cout << "Not valid start or goal" << endl;
109     return false;
110   }
111 }
112 };
```

There are few important things to consider:

- **Register the planner as BaseGlobalPlanner plugin**: this is done through the instruction:

```
5 PLUGINLIB_EXPORT_CLASS(RAstar_planner::RAstarPlannerROS, nav_core::BaseGlobalPlanner)
```

For this, it is necessary to include the library:

```
1 #include <pluginlib/class_list_macros.h>
```

- The constructors can be implemented with respect to the planner requirements and specification.

- **The implementation of the makePlan() method**: This method takes three parameters: `start`, `goal`, and `plan`. The `start` and `goal` are used to set the initial position and the target position, respectively. Those positions have to be converted from x and y coordinate to cell indices before passing them to RA*, because RA* work with indices. When RA* finishes its execution, the computed path will be returned. The cells in the path need to be converted to x and y coordinates before sending the path to the `move_base` global planner module which will publish it through the ROS topic `nav_msgs/Path`, which will then be received by the local planner module.

Now the global planner class is done, and you are ready for the second step, that is creating the plugin for the global planner to integrate it in the global planner module `nav_core::BaseGlobalPlanner` of the `move_base` package.

4.3.1.2 Compilation

To compile the RA* global planner library created above, it must be added (with all of its dependencies if any) to the *CMakeLists.txt*. This is the code to be added:

```
add_library(relaxed_astar_lib src/RAstar_ros.cpp)
```

Then, in a terminal, run `catkin_make` in your catkin workspace directory to generate the binary files. This will create the library file in the lib directory `~/catkin_ws/devel/lib/librelaxed_astar_lib`. Observe that "`lib`" is appended to the library name `relaxed_astar_lib` declared in the `CMakeLists.txt`.

4.3.2 Writing Your Plugin

The second step of integration is to deploy the planner class as a plugin. Basically, it is important to follow all the steps required to create a new plugin as explained in the plugin description page [17]. There are four steps:

4.3.2.1 Plugin Registration

First, you need to register your global planner class as plugin by exporting it. In order to allow a class to be dynamically loaded, it must be marked as an exported class. This is done through the special macro `PLUGINLIB_EXPORT_CLASS`. This macro can be put into any source (`.cpp`) file that composes the plugin library, but is usually put at the end of the `.cpp` file for the exported class. This was already done above in `RAstar_ros.cpp` with the instruction.

```
5 PLUGINLIB_EXPORT_CLASS(RAstar_planner::RAstarPlannerROS, nav_core::BaseGlobalPlanner)
```

This will make the class `RAstar_planner::RAstarPlannerROS` registered as plugin for `nav_core::BaseGlobalPlanner` of the `move_base`.

4.3.2.2 Plugin Description File

The second step consists in describing the plugin in a description file. The plugin description file is an `XML` file that serves to store all the important information about a plugin in a machine readable format. It contains information about the library the plugin is in, the name of the plugin, the type of the plugin, etc. In our case of global planner, you need to create a new file and save it in certain location in your package and give it a name, for example `relaxed_astar_planner_plugin.xml`. The content of the plugin description file (`relaxed_astar_planner_plugin .xml`), would look like this:

```
1 <library path="lib/librelaxed_astar_lib">
2   <class name="RAstar_planner/RAstarPlannerROS"
3   type="RAstar_planner::RAstarPlannerROS"
4   base_class_type="nav_core::BaseGlobalPlanner">
5     <description>This is RA* global planner plugin by iroboapp project.</description>
6   </class>
7 </library>
```

In the first line:

```
1 <library path="lib/librelaxed_astar_lib">
```

we specify the path to the plugin library. In this case, the path is `lib/librelaxed_astar_lib`, where lib is the folder in the directory `~/catkin_ws/devel/` (see Compilation section above).

```
2   <class name="RAstar_planner/RAstarPlannerROS"
3   type="RAstar_planner::RAstarPlannerROS"
4   base_class_type="nav_core::BaseGlobalPlanner">
```

Here, we first specify the name of the `global_planner` plugin that we will use later in `move_base` launch file as parameter that specifies the global planner to be used in `nav_core`. It is typically to use the namespace (`RAstar_planner`) followed by a slash then the name of the class (`RAstarPlannerROS`) to specify the name of plugin. If you do not specify the name, then the name will be equal to the type, which is in this case will be `RAstar_planner::RAstarPlannerROS`. It recommended to specify the name to avoid confusion.

The `type` specifies the name the class that implements the plugin which is in our case `RAstar_planner::RAstarPlannerROS`, and the `base_class_type` specifies the name the base class that implements the plugin which is in our case `nav_core::BaseGlobalPlanner`.

```
5   <description>This is RA* global planner plugin by iroboapp project.</description>
```

The `<description>` tag provides a brief description about the plugin. For a detailed description of plugin description files and their associated tags/attributes, please see the documentation in [18].

Why Do We Need This File? We need this file in addition to the code macro to allow the ROS system to automatically discover, load, and reason about plugins. The plugin description file also holds important information, like a description of the plugin, that does not fit well in the macro.

4.3.2.3 Registering Plugin with ROS Package System

In order for pluginlib to query all available plugins on a system across all ROS packages, each package must explicitly specify the plugins it exports and which package libraries contain those plugins. A plugin provider must point to its plugin description file in its `package.xml` inside the export tag block. Note, if you have other exports, they all must go in the same export field. In our RA* global planner example, the relevant lines would look as follows:

```
1   <export>
2     <nav_core plugin="${prefix}/relaxed_astar_planner_plugin.xml" />
3   </export>
```

The ${prefix}/ will automatically determine the full path to the file `relaxed astar planner plugin.xml`. For a detailed discussion of exporting a plugin, interested readers may refer to [19].

Important Note: In order for the above export command to work properly, the providing package must depend directly on the package containing the plugin interface, which is *nav_core* in the case of global planner. So, the *relaxed_astar* package must have the line below in its relaxed_astar/package.xml:

```
1   <build_depend>nav_core</build_depend>
2   <run_depend>nav_core</run_depend>
```

This will tell the compiler about the dependency on the `nav_core` package.

4.3.2.4 Querying ROS Package System for Available Plugins

One can query the ROS package system via `rospack` to see which plugins are available by any given package. For example:

```
1 $ rospack plugins —attrib=plugin nav_core
```

This will return all plugins exported from the nav_core package. Here is an example of execution:

```
1  turtlebot@turtlebot-Inspiron-N5110:~$ rospack plugins —attrib=plugin nav_core
2  rotate_recovery /opt/ros/hydro/share/rotate_recovery/rotate_plugin.xml
3  navfn /home/turtlebot/catkin_ws/src/navfn/bgp_plugin.xml
4  base_local_planner /home/turtlebot/catkin_ws/src/base_local_planner/blp_plugin.xml
5  move_slow_and_clear /opt/ros/hydro/share/move_slow_and_clear/recovery_plugin.xml
6  robot_controller /home/turtlebot/catkin_ws/src/robot_controller/global_planner_plugin.xml
7  relaxed_astar /home/turtlebot/catkin_ws/src/relaxed_astar/relaxed_astar_planner_plugin.xml
8  dwa_local_planner /opt/ros/hydro/share/dwa_local_planner/blp_plugin.xml
9  clear_costmap_recovery /opt/ros/hydro/share/clear_costmap_recovery/ccr_plugin.xml
10 carrot_planner /opt/ros/hydro/share/carrot_planner/bgp_plugin.xml
```

Observe that our plugin is now available under the package relaxed_astar and is specified in the file /home/turtlebot/catkin_ws/src/relaxed_astar/relaxed_astar_planner_plugin.xml. You can also observe the other plugins already existing in nav_core package, including carrot_planner/CarrotPlanner and navfn, which implements Dijkstra's algorithm.

Now, your plugin is ready to use.

4.3.3 Running the Plugin

There are a few steps to follow to run your planner in a robot. As an example, we will show the steps to test and run the plugin in Turtlebot robot. We tested the plugin using real Turtlebot robot with ROS Hydro version. However, it is expected to also work with Groovy (not tested). First, you need to copy the package that contains your global planner (in our case relaxed_astar) into the catkin workspace of your Turtlebot (e.g., catkin_ws). Second, you need to run catkin_make to export your plugin to your Turtlebot ROS environment.

Third, you need to make some modification to move_base configuration to specify the new planner to be used. For this, follow these steps:

1. In Hydro, go to this folder /opt/ros/hydro/share/turtlebot_navigation/launch/includes

```
$ roscd turtlebot_navigation/
$ cd launch/includes/
```

2. Open the file move_base.launch.xml (you may need sudo to open and be able to save) and add the new planner as parameters of the global planner, as follows:

```
1  ......
2  <node pkg="move_base" type="move_base" respawn="false" name="move_base" output="screen"↩
   >
3  <param name="base_global_planner" value="RAstar_planner/RAstarPlannerROS"/>
4  ....
```

Save and close the `move_base.launch.xml`. Note that the name of the planner is `RAstar_planner/RAstarPlannerROS` the same specified in `relaxed_astar_planner_plugin.xml`.
Now, you are ready to use your new planner.

3. You must now bringup your Turtlebot. You need to launch `minimal.launch`, `3dsensor.launch`, `amcl.launch.xml` and `move_base.launch.xml`. Here is an example of launch file that can be used for this purpose.

```
1  <launch>
2  <include file="$(find turtlebot_bringup)/launch/minimal.launch"></include>
3
4    <include file="$(find turtlebot_bringup)/launch/3dsensor.launch">
5      <arg name="rgb_processing" value="false" />
6      <arg name="depth_registration" value="false" />
7      <arg name="depth_processing" value="false" />
8      <arg name="scan_topic" value="/scan" />
9    </include>
10
11   <arg name="map_file" default="map_folder/your_map_file.yaml"/>
12   <node name="map_server" pkg="map_server" type="map_server" args="$(arg map_file)" />
13
14   <arg name="initial_pose_x" default="0.0"/>
15   <arg name="initial_pose_y" default="0.0"/>
16   <arg name="initial_pose_a" default="0.0"/>
17   <include file="$(find turtlebot_navigation)/launch/includes/amcl.launch.xml">
18     <arg name="initial_pose_x" value="$(arg initial_pose_x)"/>
19     <arg name="initial_pose_y" value="$(arg initial_pose_y)"/>
20     <arg name="initial_pose_a" value="$(arg initial_pose_a)"/>
21   </include>
22
23   <include file="$(find turtlebot_navigation)/launch/includes/move_base.launch.xml"/>
24
25 </launch>
```

Note that changes made in the file `move_base.launch.xml` will now be considered when you bring up your Turtlebot with this launch file.

4.4 ROS Environment Configuration

One important step before using the planners is tuning the controller parameters as they have a big impact on the performance. The controller parameters can be categorized into several groups based on what they control such as: robot configuration, goal tolerance, forward simulation, trajectory scoring, oscillation prevention, and global plan.

The robot configuration parameters are used to specify the robot acceleration information in addition to the minimum and maximum velocities allowed to the robot. We are working with the Turtlebot robot, and we used the default parameters from $turtlebot\_navigation$ package. The configuration parameters are set as follow: $acc\_lim\_x = 0.5$, $acc\_lim\_theta = 1$, $max\_vel\_x = 0.3$, $min\_vel\_x = 0.1$, $max\_vel\_theta = 1$ $min\_vel\_theta = -1$ $min\_in\_place\_vel\_theta = 0.6$.

The goal tolerance parameters define how close to the goal we can get. $xy\_goal\_tolerance$ represents the tolerance in meters in the x and y distance and should not be less than the map resolution or it will make the robot spin in place indefinitely without reaching the goal, so we set it to 0.1. $yaw\_goal\_tolerance$ represents the tolerance in radians in yaw/rotation. Setting this tolerance very small may cause the robot to oscillate near the goal. We set this parameter very high to 6.26 as we do not care about the robot orientation.

In the forward simulation category, the main parameters are: sim_time, $vx\_samples$, $vtheta\_samples$, and $controller\_frequency$. The sim_time represents the amount of time (in seconds) to forward-simulate trajectories, and we set it to 4.0. The $vx\_samples$ and $vtheta\_samples$ represent the number of samples to use when exploring the x velocity space and the theta velocity space respectively. They should be set depending on the processing power available, and we use the value recommended in ROS Web site for them. So, we set 8 to the $vx\_samples$ and 20 to $vtheta\_samples$. The $controller\_frequency$ represents the frequency at which this controller will be called. Setting this parameter a value too high can overload the CPU. Setting it to 2 works fine with our planners.

The trajectory scoring parameters are used to evaluate the possible velocities to the local planner. The three main parameters on this category are: $pdist\_scale$, $gdist\_scale$, and $occdist\_scale$. The $pdist\_scale$ represents the weight for how much the controller should stay close to the global planner path. The $gdist\_scale$ represents the weight for how much the controller should attempt to reach its goal by whatever path necessary. Increasing this parameter will give the local planner more freedom in choosing its path away from the global path. The $occdist\_scale$ represents the weight for how much the controller should attempt to avoid obstacles. Because the planners may generate paths very close to the obstacles or dynamically not feasible, we set the $pdist\_scale = 0.1$, $gdist\_scale = 0.8$ and $occdist\_scale = 0.3$ when testing our planners. Another parameter named dwa is used to specify whether to use the DWA when setting it to $true$, or use the Trajectory Rollout when setting it to $false$. We set it to $true$ because the DWA is computationally less expensive than the Trajectory Rollout.

4.5 Performance Evaluation

For the experimental evaluation study using ROS, we have used the real-world Willow Garage map (Fig. 4.3), with dimensions 584 * 526 cells and a resolution 0.1 m/cel. In this map, the white color represents the free area, the black color represents the obstacles or walls, and the gray color represents the unknown area.

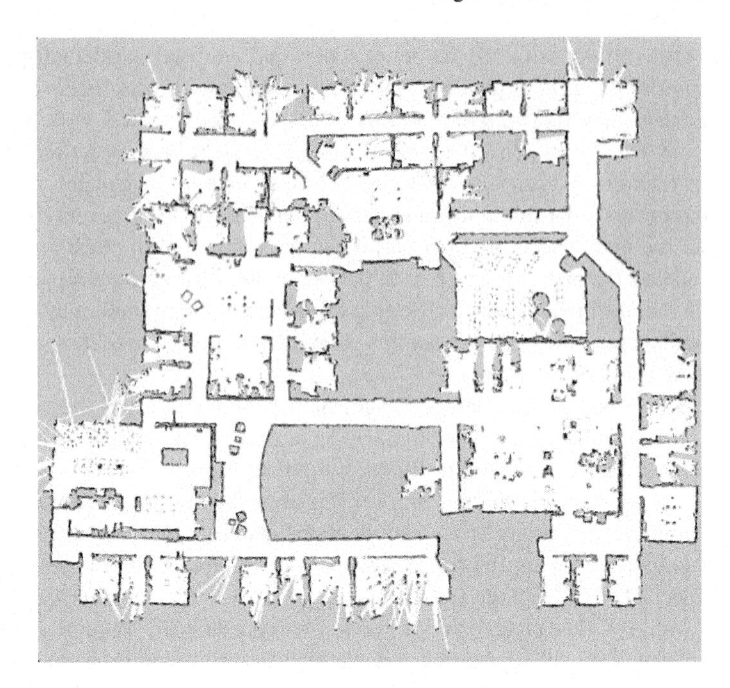

Fig. 4.3 Willow Garage map

We considered 30 different scenarios, where each scenario is specified by the coordinates of randomly chosen start and goal cells. Each scenario, with specified start/goal cells, is repeated 30 times (i.e., 30 runs for each scenario). In total, 900 runs for the Willow Garage map are performed in the performance evaluation study for each planner.

Two performance metrics are considered to evaluate the global planners: (1) *the path length*, it represents the length of the shortest global path found by the planner, (2) *the execution time*, it is the time spent by an algorithm to find its best (or optimal) solution.

Figure 4.4 shows that the RA* is faster than *navfn* in 87% of the cases. Table 4.1 shows that, in average, the RA* is much faster than *navfn*, with execution time less than the half of that of *navfn*. RA* provides near-optimal paths, which are in average only 3.08% longer than *navfn* paths. These results confirm the simulation results, presented in Sect. 4.3, about the efficiency of the RA*. Moreover, the small loss in terms of path quality does not matter in practice as this path will be used only as a guide for *local path planner*, which will generate the actual path that the robot will follow (which may be different from the global path) considering the kinematics and dynamics of the robot as well as moving obstacles, if any.

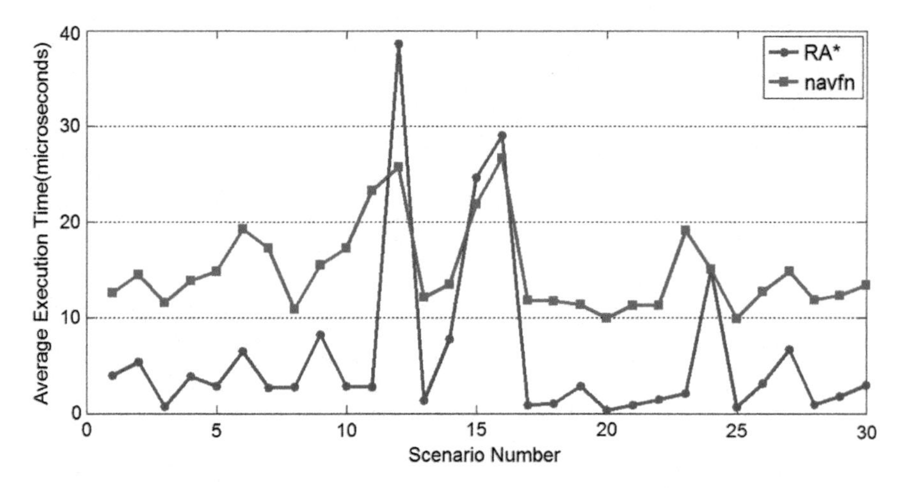

Fig. 4.4 Average execution time (microseconds) of RA* and navfn

Table 4.1 Execution time in (microseconds) and path length in (meters) of RA* and navfn

Planner	Execution time		Path length	
	Total	Average	Total	Average
RA*	185.1265156	6.170884 ± 0.504922	582.34962	19.41165
navfn	448.6488877	14.95496 ± 1.174413	564.9527	18.83176

4.6 Conclusion

In this chapter, we presented the necessary steps to integrate new path planners into the ROS global path planning component as a plugin. We demonstrated how to integrate new planner (RA*) into ROS and present its benefits. Experimentations were performed to show the effectiveness of the newly integrated planners as compared to ROS default planner.

Comparing the RA* paths with $navfn$ show that although the RA* paths are longer with small gap than the $navfn$, the execution time for RA* is 78% less than $navfn$ execution time.

References

1. Robot Operating System (ROS). http://www.ros.org.
2. Jason M. O'Kane. *A Gentle Introduction to ROS*. Independently published, October 2013. http://www.cse.sc.edu/~jokane/agitr/.

3. K.A Wyrobek, E.H. Berger, H.F.M. Van der Loos, and J.K. Salisbury. 2008. Towards a personal robotics development platform: Rationale and design of an intrinsically safe personal robot. In *2008 IEEE International conference on robotics and automation, ICRA 2008*, pages 2165–2170. IEEE.
4. Morgan Quigley, Eric Berger, and Andrew Y Ng. 2007. Stair: Hardware and software architecture. In *AAAI 2007 Robotics workshop, Vancouver, BC*, pages 31–37.
5. E. Marder-Eppstein, E. Berger, T. Foote, B. Gerkey, and K. Konolige. The office marathon: Robust navigation in an indoor office environment. In *2010 IEEE International conference on robotics and automation (ICRA)*, pages 300–307, May 2010.
6. Patrick Goebel. 2013. *ROS by example*. Lulu.
7. Cyrill Stachniss, Udo Frese, and Giorgio Grisetti. 2007. Openslam. https://openslam.org/. Accessed 29 Oct 2009.
8. Grisetti, G., C. Stachniss, and W. Burgard. 2007. Improved Techniques for grid mapping with rao-blackwellized particle filters. *IEEE Transactions on Robotics* 23 (1): 34–46.
9. Thrun, S., D. Fox, W. Burgard, and F. Dellaert. 2000. Robust Monte Carlo localization for mobile robots. *Artificial Intelligence* 128 (1–2): 99–141.
10. Anis Koubaa. 2014. Adding a global path planner as plugin in ros. http://www.iroboapp.org/index.php?title=Adding_A_Global_Path_Planner_As_Plugin_in_ROS. Accessed 27 April 2015.
11. Anis Koubaa. 2014. Writing a global path planner as plugin in ros. http://wiki.ros.org/navigation/Tutorials/Writing%20A%20Global%20Path%20Planner%20As%20Plugin%20in%20ROS.
12. Dijkstra, E.W. 1959. A note on two problems in connexion with graphs. *Numerische Mathematik* 1 (1): 269–271.
13. Hart, P.E., N.J. Nilsson, and B. Raphael. 1968. A formal basis for the heuristic determination of minimum cost paths. *IEEE Transactions on Systems Science and Cybernetics* 4 (2): 100–107.
14. Brian P Gerkey and Kurt Konolige. 2008. Planning and control in unstructured terrain. In *ICRA workshop on path planning on Costmaps*.
15. Fox, Dieter, Wolfram Burgard, and Sebastian Thrun. 1997. The dynamic window approach to collision avoidance. *IEEE Robotics & Automation Magazine* 4 (1): 23–33.
16. Anis Koubaa. 2014. Relaxed A*. https://github.com/coins-lab/relaxed_astar. Accessed 18 Feb 2015.
17. Eitan Marder-Eppstein, Tully Foote, Dirk Thomas, and Mirza Shah. 2015. pluginlib package. http://wiki.ros.org/pluginlib. Accessed 07 May 2015.
18. Mirza Shah. 2012. Plugin description file. http://wiki.ros.org/pluginlib/PluginDescriptionFile. Accessed 14 Dec 2012.
19. KenConley. 2009. Plugin export. http://wiki.ros.org/pluginlib/PluginExport. Accessed Mar 2015.

Chapter 5
Robot Path Planning Using Cloud Computing for Large Grid Maps

Abstract As discussed in Chap. 3, A^* algorithm and its variants are the main mechanisms used for grid path planning. On the other hand, with the emergence of cloud robotics, recent studies have proposed to offload heavy computation from robots to the cloud, to save robot energy and leverage abundant storage and computing resources in the cloud. In this chapter, we investigate the benefits of offloading path planning algorithms to be executed in the cloud rather than in the robot. The contribution consists in developing a vertex-centric implementation of the RA^*, a version of A^* that we developed for grid maps and that was proven to be much faster than A^* (refer to Chap. 3), using the distributed graph processing framework Giraph that rely on Hadoop. We also developed a centralized cloud-based C++ implementation of the algorithm for benchmarking and comparison purposes.

5.1 Introduction

Robots have been proved to be very successful in several applications due to their high endurance, speed, and precision. To enlarge the functional range of these robots, there is a growing need for robots to perform their computing at a large scale, which is well beyond their capabilities. These robots are usually small, lightweight, and have minimal hardware configurations and, thus, limited processing power that is insufficient for handling compute-intensive tasks. The function range of these robots can be extended only if they can offload or outsource the difficult calculations and focus on sensing and actuations. Cloud computing is a promising solution for the above problems.

In fact, the process of searching a path in large grid maps that contains different obstacles requires access to large amounts of data, which is challenging. Cloud robotics provide a very promising solution for future cloud-enabled navigation. The cloud can not only provide storage space to store the large amount of map data, but also provide processing power to facilitate the building and searching of the path quickly.

In this chapter, we used the parallelism advantages of cloud computing for robots to solve the path planning in large environments. We use the open-source graph

A. Koubaa et al., *Robot Path Planning and Cooperation*, Studies in Computational Intelligence 772, https://doi.org/10.1007/978-3-319-77042-0_5

processing framework Giraph [1] which is based on the open-source Hadoop framework [2], and we present a vertex-centric graph RA^* algorithm [3], and we used a centralized cloud-based C++ implementation of the algorithm for benchmarking and comparison purposes.

5.2 Cloud Computing and Robotics

Nowadays, we are at the cusp of a revolution in robotics. The current research trend aims at using robot systems as well as industrial robots in various applications including smart home environments [4], exploration [5], airports [6], shopping malls [7], manufacturing [8]. The International Federation of Robotics reported that the number of service robots for personal and domestic use sold in 2014 increased by 11.5% compared to 2013, augmenting sales up to USD 3.77 billion [9]. It is also forecasted that the sale of household robots could reach 25.2 million units during next three years.

Although these robots have been useful for several applications, their use is limited comparatively to other technologies such as smartphones, mobile phones, and tablets essentially for large- scale application which is well beyond their capabilities. With the emergence of cloud computing, recent studies have proposed to offload heavy computation from robots to the cloud as robots are usually small, lightweight, and have minimal hardware configurations and, thus, limited processing power that is insufficient for handling compute-intensive tasks which will not allow the robot to complete its mission in a short period of time. In fact, when connected to the cloud, robots can benefit from the powerful computational, storage, and communications resources in the cloud. This solution has a number of advantages over traditional approach, where all actions are performed on the robot. This led us to think in outsourcing the path calculation process using the cloud computing techniques in order to investigate the benefits of the cloud robotic solution to solve grid path planning. Our objective is to reduce the execution time of the path searching process in large grid maps as far as possible.

5.3 Literature Review

To the best of our knowledge, there is no previous research works that addressed the robot path planning problem in large grid maps using Giraph framework.

Some research works have tackled the shortest path problem without considering the obstacle avoidance. For example, the Giraph package [1] provides an implementation of the Dijkstra algorithm. In reference [10], the authors firstly implement the Breadth-first search algorithm using Giraph to solve the shortest path problem. They tested their algorithms in two large graphs; the first one contains 10000 vertices and 121210 edges and the second has 100000 vertices and 1659270 edges.

The authors compared the implementation of breadth-first search on Giraph with a Hadoop MapReduce solution. The simulation results confirm the superiority of Giraph over Hadoop for this kind of task. The interesting point of this paper is the proposition of new improvements that modify Giraph framework in order to support dynamics graphs, in which the input may change while the job is being performed. In paper [11], the authors described how they modify the Giraph framework to be able to support managing Facebook-scale graphs of up to one trillion edges. They presented new graph processing techniques such as composable computation and superstep splitting. Paper [12] is centered around the development of a cloud-based system, called Path Planning as a Service (PPaaS) for robot path planning. The authors proposed three-layered system architecture, which facilitates on-demand path planning software in the cloud. They have implemented the proposed system with the Rapyuta cloud engine and used ROS platform as the communication framework for the entire system.

The Okapi project [13] provides a multiple source shortest paths implementation. It effectively runs a number of single-source computations in parallel. The Okapi implementation uses a label-correcting algorithm: Each vertex stores the length of the shortest path found so far. When it is informed of a shorter path from the source vertex, it updates its stored value and informs its neighbors. Each vertex has a value of type Float and sends messages of the same type. Instead of running the single-source shortest path algorithm (SSSP) for each vertex individually, multiple sources are selected and a SSSP instance is run in parallel. This is achieved by having each vertex store a map (type MapWritable) of IDs to distances. In the same way, messages sent between vertices contain a map of source IDs to distances. For example, vertex 1 might send the message 3: 10.0, 4: 15.0 to 2, meaning that it found a new shortest path from source vertex 3, via 1, to 2 of length 10; and also a new shortest path from source vertex 4. The receiving vertex 2 then compares each new length to the values stored in its own map to decide if the path has an improvement. If at a given iteration some improvement is found, all neighbors are informed by sending the new distance. The work [14] compared the Hadoop MapReduce and the Bulk Synchronous Parallel (BSP) approach, used in Giraph, using the single-source shortest path computation, existing in the Giraph package, and the relational influence propagation algorithms. They tested the algorithm in cluster containing 85 machines, and each has 7500 MB of RAM; they used various datasets, with different sizes, to evaluate the performance of the approaches. They varied the number of nodes in the cluster and measured the execution times of the algorithm. The results revealed that iterative graph processing with BSP implementation significantly outperforms MapReduce. Pace in [15] provided a theoretical comparison of the Bulk Synchronous Parallel and the MapReduce models. In terms of graph processing they noticed that breadth-first search algorithm cannot be efficiently implemented by means of the MapReduce model. Paper [16] presents iGiraph framework for processing large-scale graphs. iGiraph is a modified version of Giraph. The authors use a new dynamic re-partitioning approach to minimize communication between computation nodes and thus reduces the cost of resource utilization. To evaluate the performance of the new framework, the authors tested the shortest path, the connected components, and the PageRank algorithms on a cloud formed by 16 machines each having 8 GB of RAM. They used three graphs containing between 403394 and 1632803. They compared the iGiraph and Giraph frameworks in terms of number of messages exchanged between partitions, the execution time and the number of workers used, and they proved that the execution time of the shortest path algorithm using iGiraph is not reduced as compared to Giraph, and it takes around 5 min. For the other algorithms, the execution time is decreased and the number of messages passing through network is reduced significantly. The authors in [17] proposed two models to maximize the performance of graph computing on heterogeneous cluster (different nodes with various bandwidths or CPU resources). They presented a greedy selection model to select the optimal worker set for executing the graph jobs for graph processing systems using hash-based partition method and a heterogeneity-aware streaming graph partitioning to balance the load among workers. Experiments were made using five different algorithms (page rank, shortest path, random walk, kcore, and weakly connected components) on two different clusters: the university laboratory cluster (46 machines) and EC2 cluster (100 instances). The experiments prove that the execution time of the algorithms is reduced by 55.9% for laboratory cluster and 44.7% for EC2 cluster as compared to traditional method. In [18], the authors compared Giraph against GraphChi. In 2012, it was proven that GraphChi is able to perform intensive graph computations in a single PC in just under 59 min, whereas the distributed systems were taking 400 min using a cluster of about 1000 computers. In this work, the authors compared the new versions of the two frameworks by testing three different algorithms (PageRank, shortest path, and weakly connected components) and they concluded that even for a moderate number of simple machines (between 20 and 40, each having 1 GB of RAM), Apache Giraph outperforms GraphChi in terms of execution time for all the algorithms and datasets used. Rathore et al. [19] proposed a graph-oriented mechanism to achieve the smart transportation system. The overall traffic information is obtained from road sensors which form big data. Graphs are generated from big data. Various graph algorithms are implemented using Giraph to achieve real-time transportation. Dijkstra is implemented to select the quickest and shortest path. In addition to the path cost, other parameters are taken into consideration to evaluate the path quality in the process of searching the shortest path such as the current traffic intensity as well as the vehicles speeds. However, experiments were conducted only on a single-node machine and they have used not very large graph (less than 90000 nodes and edges).

Table 5.1 provides more details about some works.

5.4 Hadoop: Overview

Hadoop is an open-source, Java-based framework similar to Google's Map/Reduce framework [20]. Hadoop was created by Doug Cutting and Mike Cafarella in 2006. It was originally developed to support distribution for the Nutch search engine project [21]. It provides a reliable, scalable, and distributed computing platform. Hadoop

Table 5.1 Comparison with some related works

Works	Algorithms	Simulation environment	Graph characteristics	Execution times	Key Features
[18]	PageRank, Single shortest source path (sssp) and WCC	Between 25 and 40 T2.micro Amazon Elastic Computing instances, CPU:Intel Xeon, RAM:1 GB	v = 4,847,571 / E = 68,993,773	50 seconds using 20 machines	*Performance evaluation of the algorithms using centralized and distributed graph processing systems + simulations: using different number of machines and different graphs - Why Giraph outperforms GraphChi is not explained while in 2012 it has been proven that GraphChi is more effective
[10]	Breadth-first search	Single node	V = 100,000 / E = 1,659,270	14120 milliseconds	*Performance evaluation of the Breadth-First Search algorithm + modify the giraph package to take into consideration dynamic graph - simulation: single node setup
[14]	SSSP	60 machines with 4 CPU, RAM = 7500 MB	V = 5,098,639 / E = 21,285,803	30 seconds using 60 machines	*Comparison MapReduce and Giraph using single source shortest path problem +various network datasets differing in size and structural profile -Only one metric used for performance evaluation
The single source shortest path algorithm in Giraph package	SSSP	–	–	–	*Calculate the shortest path from a fixed start node to the other nodes in the graph -Obstacle avoidance not taken into consideration
[16]	Page rank	16 machines: m1.medium NECTAR VM, DDR= 70 GB, RAM= 8 GB	V = 1,632,803 /E = 30,622,564	29000 milliseconds	*Present iGiraph a modified version of Giraph + Simulation: Evalaute the number of messages and execution time -More large graphs are not tested
[19]	Dijkstra	Single node machine, CPU = coreTMi5 machine, RAM = 4GB	V = less than 90,000/ E = not indicated	65 seconds	*Implementation of Dijkstra to solve the transportation problem +Metrics considered for evaluation: path cost, the current traffic intensity and vehicle speeds -simulation: single node machine -No implementation details provided -Larger graphs need to be tested
This work	RA^*	3 VM of the Dreamhost cloud cluster,CPU = Intel (R) Xeon(R) CPU E5-2620, RAM = 8 GB	The largest graph: V = 4,000,000 E = 7,996,000	Average: 179041,8 milliseconds	*Implementation of RA^* a path planning algorithm in grid environment +Obstacle avoidance -The available materials for experiments

supports data intensive distributed applications running on large clusters of computers. One of the main features of Hadoop is that it parallelizes data processing across many nodes (computers) in the cluster, speeding up large computations. Most of these processing occurs near the data or storage so that I/O latencies over the network are reduced. Even though Hadoop has been primarily used in search and indexing of large volumes of text files, nowadays it has even been used in other areas also like in machine learning, analytics, natural language search, and image processing. We have now found its potential application in robotics.

5.4.1 Hadoop Architecture Overview

The core of Apache Hadoop is composed of the following modules [2]:

- Hadoop Common Module: is a Hadoop base API, it contains libraries and utilities needed by other Hadoop modules;
- Hadoop Distributed File System (HDFS) (storage part): a distributed file system that stores data on commodity machines, providing very high aggregate bandwidth across the cluster;
- Yet Another Resource Negotiator (YARN): a resource management platform responsible for managing computing resources in clusters and using them for scheduling of users' applications; and
- Hadoop MapReduce (processing part): an implementation of the MapReduce programming model for large-scale data processing.

The next subsections will give a brief overview about the different modules of Hadoop.

5.4.1.1 HDFS: Hadoop Distributed File System

In [21], HDFS is defined as "a file system designed for storing very large files with streaming data access patterns, running on clusters of commodity hardware". HDFS can store files that are hundreds of megabytes, gigabytes, or terabytes in size; it is build around the idea that the most efficient data processing pattern is a write-once, read-many-times pattern. Moreover, Hadoop does not require expensive, highly reliable hardware, and it is highly fault-tolerant.

To store such a huge data, the files are stored across multiple machines. These files are stored in redundant fashion to rescue the system from possible data losses in case of failure. HDFS also makes applications available to parallel processing.

Figure 5.1 is the architecture of the Hadoop-distributed file system. HDFS has two types of nodes operating in a master worker architecture: a namenode (the master) and a number of datanodes (workers).

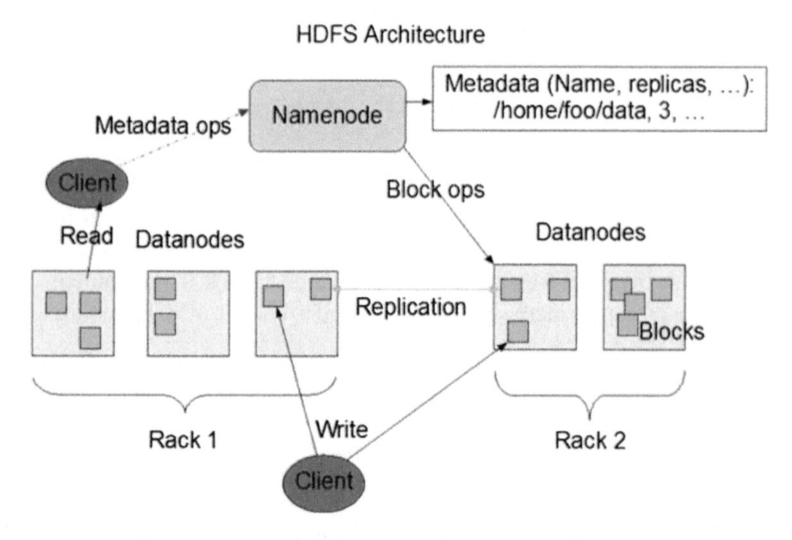

Fig. 5.1 The Hadoop distributed file system architecture

- **Namenode**: It is the master of HDFS that maintains and manages the blocks present on the DataNodes (slave nodes). It is a very high-availability server that manages the file system namespace and controls access to files by clients. In Hadoop 1.x, there is just one namenode in Hadoop which is the single point of failure in the entire Hadoop HDFS cluster. In hadoop 2.x, multiple namenode servers manage namespaces and this allows performance improvements. The namenode does the following tasks:
  - Manages the file system namespace, and it maintains the filesystem tree and the metadata for all the files and directories in the tree.
  - It directs the Datanodes (Slave nodes) to execute the low-level I/O operations.
  - It keeps a record of how the files in HDFS are divided into blocks.
  - The NameNode is also responsible to take care of the replication factor of all the blocks. If there is a change in the replication factor of any of the blocks, the NameNode will record this in the EditLog.
  - Regulates client's access to files: it provides information to the clients about how the file data is distributed across the data nodes.
  - It also executes file system namespaces operations such as renaming, closing, and opening files and directories.
- **Datanodes**: are the slave nodes in HDFS; a datanode is a commodity hardware, that is, a non-expensive system which is not of high quality or high availability. For every node in a cluster, there will be a datanode which manages storage attached to the node that it runS on. The datanodes are responsible of the following tasks:
  - Datanodes perform read-write operations on the file systems.

  - They also perform operations such as block creation, deletion, and replication according to the instructions of the namenode.
  - They regularly send a report on all the blocks present in the cluster to the NameNode.

- **Block**: Generally, the user data is stored in the files of HDFS. The file in a file system will be divided into one or more segments (chunks) and/or stored in individual data nodes. These file segments are called as blocks. In other words, the minimum amount of data that HDFS can read or write is called a block. The default block size is 128 MB, but it can be increased as per the need to change in HDFS configuration.

5.4.1.2 Map/Reduce Framework: Overview

One of the main functionality in Hadoop is the MapReduce framework which is described in [22]; it has been first developed by Google [20]. The MapReduce is a programming model for processing and generating large datasets with a parallel, distributed algorithm on a cluster inside the HDFS. It provides a mechanism for executing several computational tasks in parallel on multiple nodes on a huge dataset. This reduces the processing or execution time of computationally intensive tasks by several orders of magnitude compared to running the same task on a single server. Figure 5.2 illustrates how the MapReduce model works. MapReduce works by breaking the processing into two phases: the map phase (map task) and the reduce phase (reduce task). A MapReduce job usually splits the input dataset into independent chunks (blocks) which are processed by the map tasks in a completely parallel manner. The framework sorts the outputs of the maps, which are then input to the reduce tasks. Typically, both the input and the output of the job are stored in a file system. The framework takes care of scheduling tasks, monitoring them, and re-executes the failed tasks.

5.4.1.3 Yet Another Resource Negotiator :YARN

In the second version of Hadoop, a new module appeared and changed the architecture of Hadoop significantly called YARN. It is responsible for managing cluster resources and job scheduling. In Hadoop 1.x, this functionality was integrated with the MapReduce module where it was realized by the JobTracker component. The fundamental idea of YARN is to split the two major functionalities of the JobTracker, resource management, and task scheduling/monitoring in order to have a global ResourceManager and an ApplicationMaster. The YARN module consists of three main components: *(1)* Global Resource Manager (RM) per cluster, *(2)* Node Manager (NM) per node and *(3)* Application Master (AM) per job

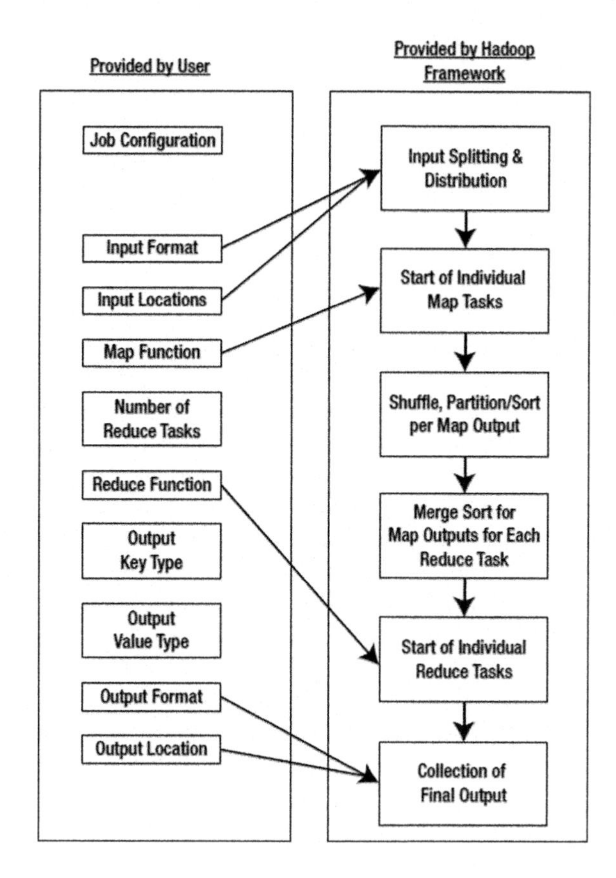

Fig. 5.2 Parts of a MapReduce job

- Resource Manager is the core component of YARN (the master). It knows where the slaves are located, it manages the resources and makes allocation decisions. It runs several services, the most important are the scheduler and the Applications Manager.
- Node Manager (many per cluster) is the slave of the infrastructure. It is responsible for launching containers (a fraction of the NodeManager capacity and it is used by the client for running a program), each of which can house a map or reduce task. For each node in the cluster is assigned a NodeManager. The role of the NodeManager consists to keep up to date with the ResourceManager, overseeing containers' life-cycle management; monitoring resource usage (memory, CPU) of individual containers, tracking node-health and log's management.
- ApplicationMaster is responsible for the execution of a single application. It asks for containers to the Resource Scheduler (Resource Manager) and executes specific programs (e.g., the main of a Java class) on the obtained containers.

5.5 Giraph: Overview

Hadoop provides parallel computing platform with fault tolerance support for big data-related problems. However, MapReduce programming model does not suit well for graph-related problems. In MapReduce programming model, for each job, output is written to disk, and for new job, graph is again loaded into the memory. This loading of graph, every time a job is started, is an overhead for large graphs. To avoid such overheads, a new parallel computing platform GIRAPH [1] is introduced by Apache in 2012, which suits well for graph applications. Apache Giraph is open source. It is a loose implementation of Google's Pregel [23]which is inspired by the Bulk Synchronous Parallel (BSP). Both Pregel and Giraph employ a vertex-centric (think-like-a-vertex) programming model to support iterative graph computation. Giraph adds several features beyond the Pregel model, including master computation, sharded aggregators, edge-oriented input, out-of-core computation, and more. Giraph can run as a typical Hadoop job that uses the Hadoop cluster infrastructure. In Giraph, each vertex is a single computation unit, which contains its internal state (identifier and value) and all outgoing edges. Thus, the abstraction is made on a vertex-centric level, which is more intuitive for graph algorithms. The graph-processing programs are expressed as a sequence of iterations called supersteps. During a superstep, the framework starts a user-defined function for each vertex, conceptually in parallel. The user-defined function specifies the behavior at a single vertex and a single superstep S. The function can read messages that are sent to the vertex in superstep $S - 1$, send messages to other vertices that will be received at superstep $S + 1$, and modify the state of the vertex and its outgoing edges. Messages are typically sent along outgoing edges, but you can send a message to any vertex with a known identifier.

5.5.1 *Giraph Architecture*

Giraph applies a master/worker architecture, illustrated in Fig. 5.3. During program execution, graph is loaded and partitioned by the master and assigned to workers. The default partition mechanism is hash partitioning, but custom partition is also supported. The master then dictates when workers should start computing consecutive supersteps. The master is responsible also about the coordination and synchronization it requests checkpoints and collects health statuses. Like Hadoop, Giraph uses Apache ZooKeeper for synchronization. Workers are responsible for vertices. A worker starts the compute function for the active vertices. It also sends, receives, and assigns messages with other vertices. During execution, if a worker receives input that is not for its vertices, it passes it along. Once the computation has halted, workers save the output. Checkpoints are initiated at user-defined intervals and are used for automatic application restarts when any worker fails. Any worker can act as the master and one will automatically take over if the current master fails.

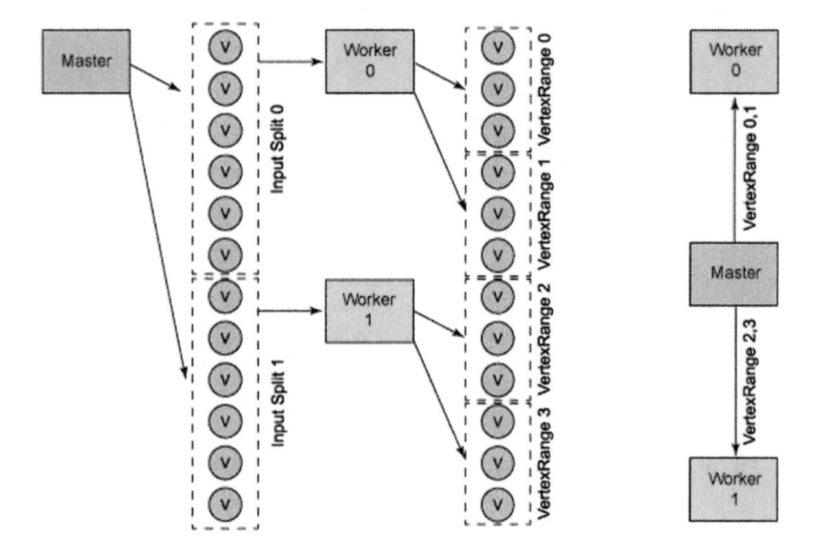

Fig. 5.3 The Giraph Architecture

5.5.2 *The Bulk Synchronous Parallel Model*

A Bulk Synchronous Parallel (BSP) computation model consists of a series of global supersteps or iterations. Each superstep consists of three components as depicted in Fig. 5.4:

1. Concurrent computation: each processor is assigned a number of vertices and processes in parallel. Every participating processor may perform local computations, i.e., each process can only make use of values stored in the local fast memory of the processor. The computations occur asynchronously of all the others but may overlap with communication.
2. communication: the processors exchange messages between each other;
3. barrier synchronization: when a processor reaches the barrier, it waits until all other processors finish their communications.

According to BSP, we have n processing units, which can communicate through a medium such as a network or a bus. The input is divided across the processing units, and each processing unit computes its intermediate results to its subproblem locally. When the processing units have finished, they exchange the intermediate results according to the semantics of the algorithm. When a processing unit has finished computing its subproblem and sending its intermediate results, it waits for the others to finish as well. When all processing units have finished, they go with the next superstep, computing their subproblems based on their previously computed state and the messages that have received. The waiting phase is the synchronization barrier.

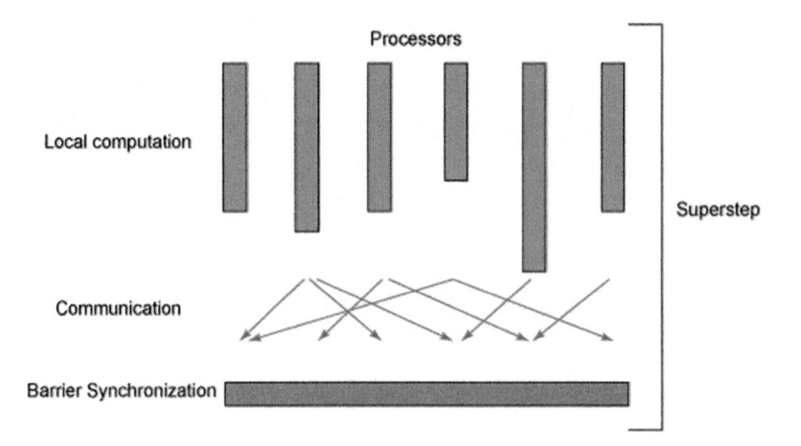

Fig. 5.4 The BSP Model

5.6 Implementation of RA^* Using Giraph

The RA^* algorithm is a version of A^* that has been developed for grid maps and that was proven to be much faster than A^* in Chap. 3. This section presents a vertex-centric RA^* algorithm designed using Giraph and Hadoop frameworks for solving the global path planning problem in large grid environments.

As mentioned in the Sect. 5.5, when designing a Giraph algorithm, we should consider the "think like-a-vertex" programming model, i.e., the algorithm should be transformed to show the behavior of one vertex in a given superstep.

Step 1: Input Format Selection:

As explained in the previous section, the first step of any Giraph application is to choose a MapReduce Input Format. For the RA^* job, we choose the `Json Long Double Float Double Vertex Input Format` format to present the grid map as a graph. The grid will be transformed to a series of lines that represent the graph (the input file of the job); each line has the following format:

$JSONArray(vertexID, vertexValue, JSONArray(JSONArray$ $(DestVertexID, EdgeValue), ...))$

Each line is composed of three elements to describe one cell in the grid. The $vertexID$ is the cell ID, $vertexValue$ is the cell value which can have one of these two possible values 0 or 100, where 0 means the cell is free and 100 means the cell is occupied by an obstacle and finally a list of adjacent neighbors, each neighbor is represented by its ID $DestVertexID$ and the edge value $EdgeValue$. The edge value has two possibles value 1 or 1.4; it is equal to 1 if the neighbor vertex exists in the vertical or horizontal direction in the map and 1.4 if the adjacent vertex exists in the diagonal direction.

The following example represents a 10*10 grid map represented using the JsonLongDoubleFloatDoubleVertexInputFormat (only the first three vertices are presented, the whole file contains 99 vertices):

```
[0,0,[[1,1],[10,1],[11,1.4]]]

[1,0,[[0,1],[2,1],[10,1.4],[11,1.4],[12,1]]]

[2,0,[[1,1],[3,1],[11,1.4],[12,1.4],[13,1.4]]]

[3,0,[[2,1],[4,1],[12,1.4],[13,1.4],[14,1.4]]]

....
```

Step 2: Implementation of the Compute Method:

The RA^* algorithm is implemented as a subclass of the `AbstractComputation` class that provides a `compute()` method which will be executed by each *active* vertex in the graph in each superstep. Vertices can have two different states during the execution of the algorithm: actives or inactives. They all begin the process as actives, then they become inactive if they vote to halt. When they are inactive, they do not execute the compute method, and they can come back to activity only if they receive a message from neighbors. At the end of the superstep, it is ensured that all active vertices have executed the compute method and messages are being delivered for the next superstep.

This method has two inputs as shown in Algorithm 1, the vertex expanded $Vertex < LongWritable, DoubleWritable, FloatWritable >$ and the message received by this vertex. In our algorithm, the JsonLongDoubleFloatDoubleVertexInputFormat is used to represent the vertices and edges of the graph, so the types defined in the vertex definition aligns with the input format.

Algorithm 1: The compute method signature

```
1 @Override
2 public void compute (Vertex < LongWritable, DoubleWritable,FloatWritable > vertex,
  Iterable < DoubleWritable > messages)
3 {
4 ...
5 }
```

The same compute method is called for all iterations, so any special cases that need to be handled in a particular superstep must be done through a call to the `getSuperstep()` function. For example, in superstep 0, all vertices are actives at the beginning. However, only the start vertex updates its value or f_score and sends messages to its neighbor's vertices. Other free vertices set their values to the

Algorithm 2: Superstep 0 of The Algorithm

```
   // if we are in superstep 0 and the processed vertex is the
   start vertex
1  if (getSuperstep() == 0) then
2      if (isStart(vertex)) then
           // update the h_score value of the start vertex
3          h_score=calculateHScoreValue();
           // send message containing its g_score value to the
           neighbors
4          sendCurrentVertexMessages(vertex);
           // update the f_score of the vertex
5          vertex.setValue(h_score);
6      else
7          vertex.setValue(MAX_VALUE));
8      end
9      vertex.voteToHalt();
10 end
```

maximum double value. At the end of superstep 0, all vertices vote to halt and become inactives. This is presented in Algorithm 2.

In the next supersteps, an active vertex that invokes the compute method can: ***(1)*** **Send messages to its adjacent nodes**, if it is elected to be the next current vertex, in order to update the g_score values of the neighbors. The `sendMessage()` implemented within the Giraph package enables the vertex to send only its value to its neighbors; thus, we implemented our own `sendMessages()` method (Algorithm 3) in order to send both the ID and the value (g_score) of the current vertex to the neighbors. ***(2)*** **Receive a message from its neighbors**: (Algorithm 4) neighbors receive the current vertex ID and g_score in order to update their g_score and f_score. ***(3)*** **Make computation**: (Algorithm 5) the vertex that receives message updates its value (g_score) then it will be added to the open list.

Algorithm 3: Send Message Method

```
1 public void sendCurrentVertexMessages(Vertex) {
2 for each edge of the current vertex do
3     sendMessage(DestVertexID, Message(SenderVertexID,SenderVertexGScore));
4 end
5 }
```

Finally, after sending all messages, the vertex will vote to halt processing using the method `voteToHalt()`. The overall execution will be halted when the target vertex is found or the master stops the computation using `haltComputation()` method if the path has not been found.

Algorithm 4: Receive Message Method

```
1 for (each received message) do
2   CurrentVertexGScore=getGScoreOfSenderVertex (message);
3   CurrentVertexID=getIDOfSenderVertex (message);
4 end
```

Algorithm 5: Vertex Computation

```
1 g_score=CurrentVertexGScore+dist_edge (VertexID, CurrentVertexID);
2 h_score=calculateHScoreValue (VertexID, CurrentVertexID);
3 vertex.setValue(g_score+h_score);
4 AddToOpenList(vertexID, f_score);
5 aggregate(OPENLIST_AGG, newElementAddedToOPL);
```

Step 3: Implementation of the RAStar Master Compute class:

The RA^* Master Compute `RAStarMasterCompute` is a subclass of `Default Master Compute` class, it is a way to introduce centralization into our algorithm. It executes some computation between supersteps. In each superstep, the RAStar-MasterCompute runs first then the workers execute the compute method for their vertices. Before each superstep, the RAStarMasterCompute is called to register the aggregators using the `initialize()` function (this will be explained in detail in the next paragraph), then the `compute()` method of the RAStarMasterCompute class is invoked; in this method, we choose the vertex that has the minimum f_score existing in the open list to be the next current vertex. The new current is removed from the open list and shared among all the workers. If the open list is empty and the target is not found, then the algorithm computation is stopped using the function `haltComputation()`; otherwise, the search process continues until reaching the goal cell if a path exists.

Algorithm 6: RAStarMasterCompute class: compute method

```
1 if (getSuperstep()>1) then
2
3   if (openList is not empty && goalVertex is not found) then
4     currentVertex = the node in openList having the lowest f_score;
5     newOpenList=remove currentVertex from openList;
6   end
    // add the current vertex and the new open list to the
    aggregator to be shared between workers;
7   Add currentVertex to the current vertex aggregator;
8   Add newOpenList to the openList aggregator;
9 end
```

Aggregators They are global objects visible to all vertices; they are used for coordination and data sharing. Each aggregator is assigned to one worker that receives the partial values from the other workers and is responsible for performing the aggregation. Afterward, it sends the final value to the master. Moreover, the worker in charge of an aggregator sends the final value to all the other workers.

During supersteps, vertices provide values to the aggregator. These values will be available for other vertices in the next superstep. Different types of aggregators are used in our algorithm. We use an aggregator for the openList `MyDouble Dense Vector Sum Aggregator`, each expanded vertex is added to the openList if it does not exist. Another aggregator is used to indicate if the goal vertex is found or not `BooleanOverwriteAggregator`. Providing a value to aggregator is done by calling the `aggregate()` function. To get the value of an aggregator during the previous supestep, we used `getAggregatedValue()` function. The aggregators are registered in the RAStarMasterCompute class in *initialize*() fuction by calling `registerAggregator()` function or `registerPersistent Aggregator()` according to the aggregator type chosen regular or persistent. The value of a regular aggregator will be reset to the initial value in each superstep, whereas the value of persistent aggregator will live through the application.

Many aggregators are already implemented in the giraph package. In RA^* job, we implemented three new aggregators. The new aggregator extends `Basic Aggregator` class, two functions must be implemented the `aggregate()` function and the `createInitialValue()` function as describing the example in Algorithm 7:

Algorithm 7: LongDenseVectorOverwrite Aggregator

```
1 @Override
2 public LongDenseVector createInitialValue() {
3 return new LongDenseVector();
4 }
5
6 @Override
7 public void aggregate(LongDenseVector vector) { if (vector.length()!=0) then
8 |  getAggregatedValue().overwrite(vector);
9 end
10 }
```

5.7 Performance Evaluation

In this section, we present the results of various experiments carried out to investigate the performance of the RA^* algorithm implemented using Giraph framework. We compared the results of the vertex-centric RA^* algorithm against the C++ version.

Table 5.2 Grid Maps Characteristics

Map size	Number of vertices	Number of edges
500*500	250000	499000
1000*1000	1000000	1998000
2000*2000	4000000	7996000

5.7.1 *Cloud Framework*

We tested our algorithms on a small Dreamhost cloud cluster [24] formed by seven virtual machines, one Master and six slaves, each with three 2.10 GHz processors Intel (R) Xeon (R) CPU E5-2620, 8GB RAM and a 80 GB disk, running Ubuntu 14.4 as an operating system. The master node runs only the namenode task, whereas all computations are performed in the datanodes running in the slaves machines. We used Hadoop 2.4.0 and Giraph 1.2.0.

5.7.2 *Experimental Scenarios*

We considered three different maps for test: The first with dimensions 500*500 cells, the second is of size 1000*1000, and the third map is 2000*2000 map. Table 5.2 provides details about the maps that we have used for simulation. We considered 21 different scenarios to test the three maps, where each scenario is specified by the coordinates of randomly chosen start and goal cells. Each scenario, is repeated five times (i.e., five runs for each scenario) and with six different numbers of workers. In total, 630 runs to evaluate the performance of the vertex-centric RA^*. For each run, we recorded the length of the generated path and the execution time of RA^* (without Hadoop initialisation time) and the time required by Hadoop to initialise the job. The execution time of an algorithm in a given map is the average of the five execution times recorded, calculated with 95% of confidence interval.

In what concerns paths found, the paths generated by the two implementation of RA^* (vertex-centric and C++) are the same of all maps and for all runs.

5.7.3 *Impact of Number of Workers*

In this section, we investigate the impact of the number of workers used by Giraph on the execution time of the vertex-centric RA^* algorithm. We vary the number of workers used by Giraph to execute the RA^* job, and we examine its effect on the execution time. For each experiment, we fix the number of workers and we perform five runs of the algorithm, and we record the execution time.

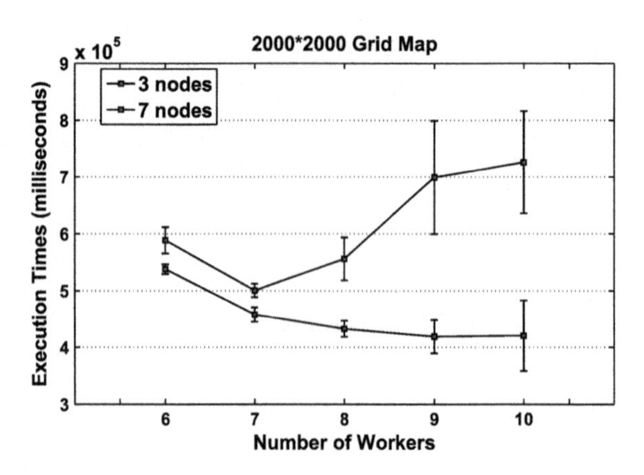

Fig. 5.5 Average execution times of RA* implemented using Giraph/Hadoop for the different grid maps

Figure 5.5 shows that, with three machines in the cluster, the best number of workers for 500*500 grid map is three, whereas larger maps need large number of workers, for 1000*1000, five workers are needed and for 2000*2000 the best number of workers is seven. We clearly see that with more numbers of nodes in the cluster (seven machines), the best number of workers increases for all maps, it is five for 500*500, seven for 1000*1000, and seven for 2000*2000 grid maps. For all maps, using more or less than needed workers for the graph increases the execution times. In the case of more number of workers, this is explained by the increasing of the execution time needed for communication between the workers which contributes in increasing the execution time of the whole algorithm, whereas in the case of less number of workers, the increasing of the algorithm runtime is related to the increasing number of vertices handled by each worker.

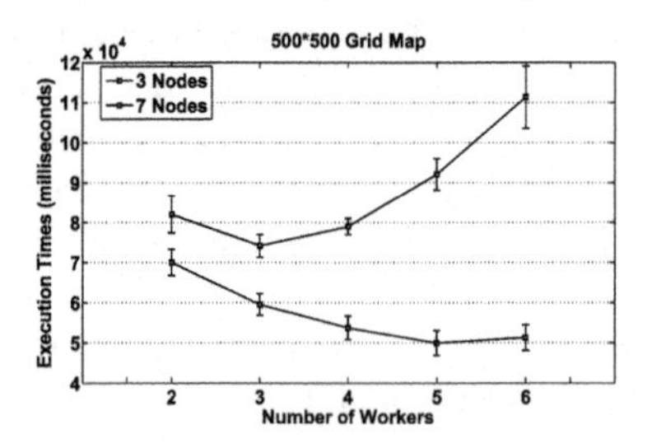

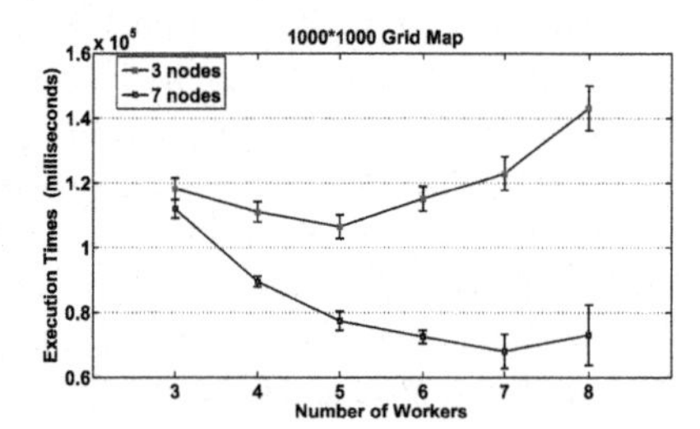

5.7.4 Execution Times

Figure 5.6 compares the average execution times consumed by the RA^* implemented using Giraph framework, the RA^* using C++ and the Hadoop initialization. We clearly observe from these figures, that the difference in execution time between the two algorithms is significant, for example for 1000*1000 grid map RA^* implemented using Giraph and executed in a cluster composed from three machines is seventy

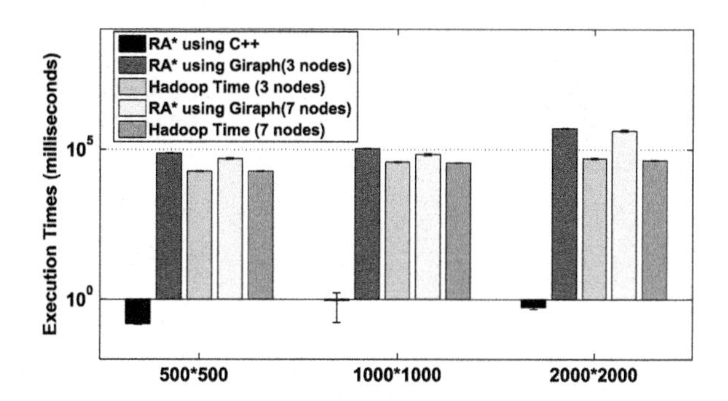

Fig. 5.6 Average execution times of the different implementation of RA* and Hadoop initialisation time for 500*500, 1000*1000 and 2000*2000 grid maps

thousand times larger than RA^* implemented using C++. This can be explained by different reasons; first, the numbers of iterations of RA^* implemented using Giraph is higher than RA* implemented on C++. This is related to the following reasons: The first reason is the Giraph programming concepts. In fact, in superstep i, the current vertex sends a message to their neighbors in order to update their g_score values, the messages will be received in the next superstep (i+1), thus to update the g_score we need two superstep one for sending messages, one for updating g_score of neighbors which contributes in increasing the number of superteps and then the execution time of the whole algorithm. Moreover, the openList in the algorithm cannot be implemented only as an aggregator as it must be shared between all workers. The aggregators are manipulated only by the master, it can add/remove values to/from the openList, so if the neighbors of the current vertex are added to the openList in superstep i, this will be visible for workers only in superstep i+1 after the master update which also contributes in increasing the number of supersteps and then the execution time of the algorithm. In addition, we should note that the time consumed by the RA^*, it not only the time required to do computation but also communication between workers which also contribute in the increasing of the run time.

In addition, we see that the increasing of the number of machines in the cluster from three to seven contributes in reducing the execution times; for 500*500 and 1000*1000 grid map the runtime is reduced up to 34,2% and for 2000*2000 grid map up to 16.2%. However, the C++ version of RA^* always exhibits the best execution time.

Also, looking to Fig. 5.6, we observe that the time required by Hadoop for the initialisation and the shutdown of the job, without considering the computation time, is greater than the runtime of the RA^* implemented using C++. For example, for 1000*1000 grid map, the Hadoop initialisation time is twenty thousand higher than RA^* C++ version. This clearly proves that Giraph/Hadoop framework (using two slaves each has 8GB of RAM) is not an appropriate technique for solving the path planning problem in large environments. But this does not mean that cloud computing techniques are not efficient for solving the large-scale path planning problem, with a cluster containing more than two slaves that are more powerful, may be we can achieve good results. As shown in Fig. 5.7, when we increase the RAM size from 4 to 8GB, we reduced the execution time of the RA^* up to 27,6%.

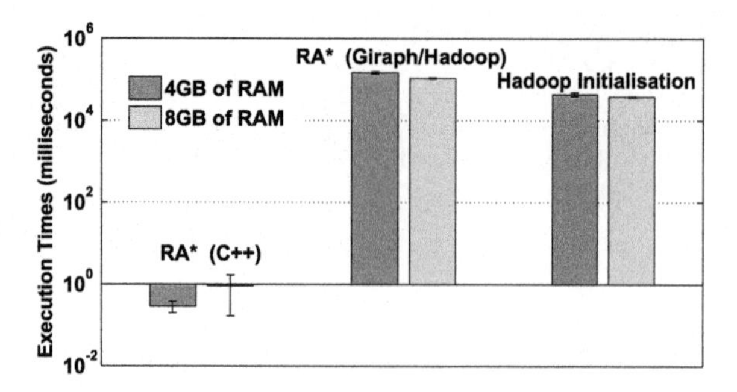

Fig. 5.7 Average execution times of RA* implemented using Giraph/Hadoop for 1000*1000 grid map tested for different RAM sizes

5.7.5 *Total Number of Messages Exchanged, Memory Footprint and CPU Usage of RA^**

To count the number of messages exchanged for RA^*, we fix the number of workers to the best number that reduces the execution time, and we perform five runs of the algorithm for each randomly chosen scenario, and we record the number of messages transferred between vertices.

5.7.5.1 Total Number of Messages

We should distinguish two different types of messages: Local messages between vertices (or subgraphs) belonging to the same worker co-located on the same machine, and remote messages transferred over network between vertices (subgraphs) on different machines. In fact, before the computation phase vertices are hashed and distributed across multiple machines. To exchange their status, the vertices send messages. When the vertex sends a message, the worker first determines whether the destination vertex of the message is owned by a remote worker or the local worker. Figure 5.8 shows the total number of messages exchanged between vertices of the graph for RA^*. We should note that a smaller number of messages does not

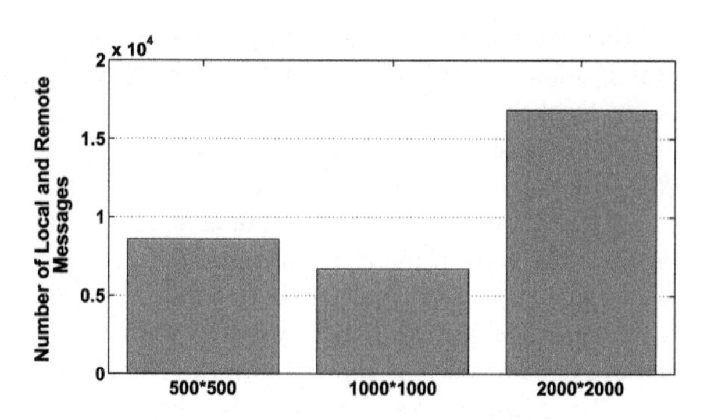

Fig. 5.8 Number of messages (local and remote) exchanged of RA^* for different grid maps

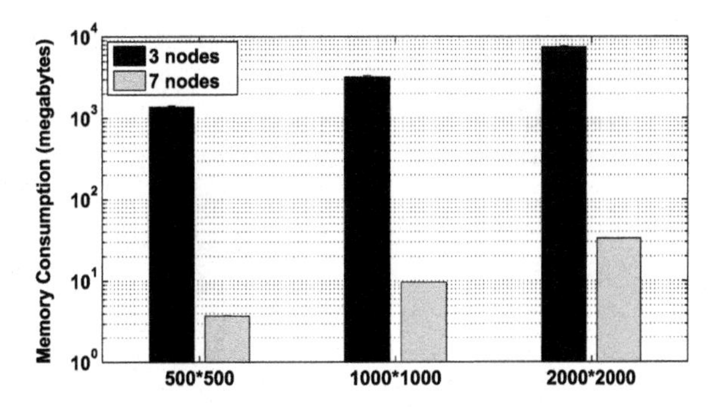

Fig. 5.9 Memory consumption of RA^* implemented using Giraph/Hadoop and RA^* implemented using C++ for different grid maps

always translate a faster execution time. For example, the number of messages sent for 1000*1000 grid map is less than 500*500 grid map, but the time needed to execute the scenarios of 1000*1000 grid map is greater than 500*500 grid map. This is explained by the distribution of vertices between workers (partitions) and the scenario (start/goal) position (near or far).

5.7.5.2 Memory Footprint

Figure 5.9 illustrates the memory consumption of the RA^* implementations for the different grid maps tested. We observe from this figure that the Giraph/Hadoop implementation requires 99% more memory than C++ implementation. There are several possible reasons for this behavior: In fact, the memory depends on the underlying programming language, Giraph uses the Java programming language which requires more memory than C++. Moreover, the framework structure of Giraph also contributes in increasing the memory usage. For example, the incoming messages sent in a vertex program are cached and only sent and treated at the end of a superstep. Also, the states of vertices that are modified at a given superstep are stored to become available only on the next superstep. All these factors will contribute in increasing the memory consumption of the Giraph implementation.

5.7.5.3 CPU Usage

Figure 5.10 presents the CPU usage of the two version of RA^* algorithm. We observe that the CPU time of the algorithms grow as the input vertices increase. Also, the CPU usage of the Giraph implementation is higher than the C++ implementation, for example for 500*500 grid map the CPU usage of the RA^* Giraph version is 12 thousands times more than the C++ implementation.

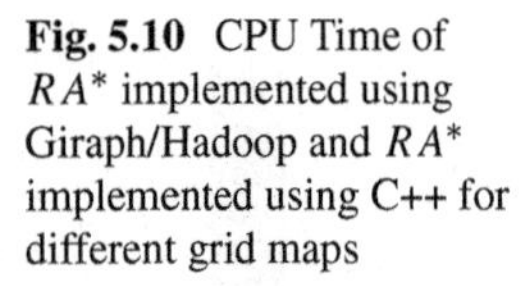

Fig. 5.10 CPU Time of RA^* implemented using Giraph/Hadoop and RA^* implemented using C++ for different grid maps

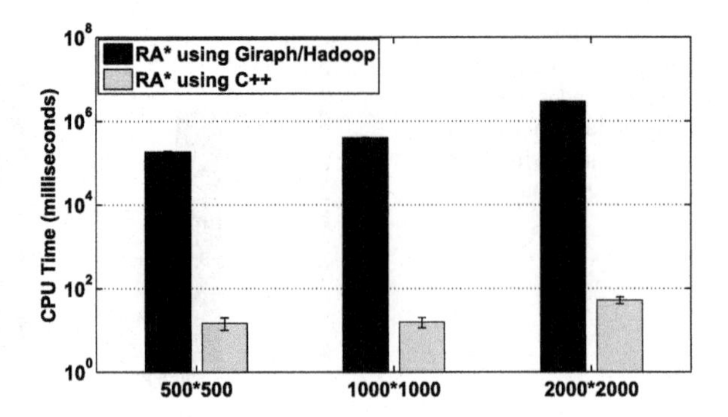

5.8 Lessons Learned

We have thoroughly analysed and compared two different versions of a relaxed version of A^* called RA^* implemented to solve the path planning problem in large grid maps; a vertex-centric version of RA^* is implemented using cloud computing techniques: The open source-distributed graph processing framework Giraph that works on the top of Hadoop and a centralized version is implemented using C++. We retain the following general lessons from the performance evaluation presented in the previous sections:

- **Lesson 1**: Experiments demonstrate that cloud computing techniques are not efficient for the robot global path planning problem using the available machines (seven machines each having three 2.10 GHz processors Intel (R) Xeon (R) CPU E5-2620, 8GB RAM and a 80 GB disk. In fact, it has been proved that the centralized C++ implementation of RA^* outperforms the vertex-centric version in terms of execution time, memory consumption and CPU time.
  This is related to the following reasons: *(1)* the Giraph programming concepts (messages sending, the aggregators used to implement the openList) which contributes in increasing the number of superteps (iterations) of the algorithm and then the execution time of the whole algorithm. *(2)* The execution time needed to execute the algorithm is not composed only from the computation time but also the communication time between workers that manipulate different partitions of the graph; this also will increase the runtime of the Giraph/Hadoop implementation of RA^*. *(3)* The Hadoop time which is required for setup, initialization, and shutdown (without considering the computation time) is greater than the C++ implementation.
- **Lesson 2**: Increasing the number of nodes in the cluster from three to seven reduces the execution time. However, RA^* C++ implementation always outperforms the vertex-centric version. This proved that Giraph graph processing framework and Hadoop are not efficient in solving path planning problem for large grid maps. Other cloud techniques with more powerful machines used for experiments might be efficient. This will be investigated in the future.

5.9 Conclusion

In this chapter, we implemented a vertex-centric RA^* using the open-source graph processing framework Giraph that works on the top of Hadoop framework. We compared the efficiency of the new version of RA^* with a C++ implementation of RA^*. Our objective was to reduce the execution time of the algorithm as far as possible to solve the global robot path planning in large-scale grid maps. Experiments show that cloud computing techniques are not efficient for the robot global path planning problem using the available materials. However, our experiments was performed in a small cluster containing only two slaves each has 8GB of RAM, more better results could be achieved if we use more performant cluster (more number of slaves).

References

1. Giraph. 2012. http://giraph.apache.org/. Accessed 2016.
2. Hadoop. 2008. http://hadoop.apache.org/. Accessed 2017.
3. Ammar, Adel, Hachemi Bennaceur, Imen Châari, Anis Koubâa, and Maram Alajlan. 2016. Relaxed dijkstra and a* with linear complexity for robot path planning problems in large-scale grid environments. *Soft Computing* 20 (10): 4149–4171.
4. Mustafa Al-Khawaldeh, Ibrahim Al-Naimi, Xi Chen, and Philip Moore. 2016. Ubiquitous robotics for knowledge-based auto-configuration system within smart home environment. In *Information and Communication Systems (ICICS), 2016 7th International Conference on*, 139–144. IEEE.
5. Nobutaka Tanishima, Hiroki Kato, Toshimichi Tsumaki, and Keiichi Yanagase. 2017. Rocket-propelled exploration robot: Shooting scouter, concept and evaluation of flight dynamics. In *Aerospace Conference, 2017 IEEE*, 1–7. IEEE.
6. Jie-Hua Zhou, Ji-Qiang Zhou, Yong-Sheng Zheng, and Bin Kong. 2016. Research on path planning algorithm of intelligent mowing robot used in large airport lawn. In *Information System and Artificial Intelligence (ISAI), 2016 International Conference on*, 375–379. IEEE.
7. Takayuki Kanda, Masahiro Shiomi, Zenta Miyashita, Hiroshi Ishiguro, and Norihiro Hagita. 2009. An affective guide robot in a shopping mall. In *Human-Robot Interaction (HRI), 2009 4th ACM/IEEE International Conference on*, 173–180. IEEE.
8. Chen, Chiu-Hung, Tung-Kuan Liu, and Jyh-Horng Chou. 2014. A novel crowding genetic algorithm and its applications to manufacturing robots. *IEEE Transactions on Industrial Informatics* 10 (3): 1705–1716.
9. The international federation of robotics. 2017. http://www.ifr.org/service-robots/statistics/. Accessed 2017.
10. Marco Aurelio, Barbosa Fagnani, and Gomes Lotz. 2013. *Dynamic Graph Computations using Parallel Distributed Computing Solutions*. PhD thesis, Queen Mary, University of London.
11. Ching, Avery, Sergey Edunov, Maja Kabiljo, Dionysios Logothetis, and Sambavi Muthukrishnan. 2015. One trillion edges: Graph processing at facebook-scale. *Proceedings of the VLDB Endowment* 8 (12): 1804–1815.
12. Miu-Ling Lam and Kit-Yung Lam. 2014. Path planning as a service ppaas: Cloud-based robotic path planning. In *Robotics and Biomimetics (ROBIO), 2014 IEEE International Conference on*, 1839–1844. IEEE.
13. The Okapi Project. 2012. https://github.com/grafos-ml/okapi. Accessed 2015.

14. Tomasz Kajdanowicz, Wojciech Indyk, Przemyslaw Kazienko, and Jakub Kukul. 2012. Comparison of the efficiency of mapreduce and bulk synchronous parallel approaches to large network processing. In *Data Mining Workshops (ICDMW), 2012 IEEE 12th International Conference on*, 218–225. IEEE.
15. Matthew Felice Pace. 2012. Bsp vs mapreduce. *Procedia Computer Science* 9: 246–255.
16. Safiollah Heidari, Rodrigo N Calheiros, and Rajkumar Buyya. 2016. iGiraph: A cost-efficient framework for processing large-scale graphs on public clouds. In *Cluster, Cloud and Grid Computing (CCGrid), 2016 16th IEEE/ACM International Symposium on*, 301–310. IEEE.
17. Jilong Xue, Zhi Yang, Shian Hou, and Yafei Dai. 2015. When computing meets heterogeneous cluster: Workload assignment in graph computation. In *Big Data (Big Data), 2015 IEEE International Conference on*, 154–163. IEEE.
18. Junnan Lu and Alex Thomo. 2016. An experimental evaluation of giraph and graphchi. In *Advances in Social Networks Analysis and Mining (ASONAM), 2016 IEEE/ACM International Conference on*, 993–996. IEEE.
19. M. Mazhar Rathore, Awais Ahmad, Anand Paul, and Gwanggil Jeon. 2015. Efficient graph-oriented smart transportation using internet of things generated big data. In *Signal-Image Technology & Internet-Based Systems (SITIS), 2015 11th International Conference on*, 512–519. IEEE.
20. Dean, Jeffrey, and Sanjay Ghemawat. 2008. Mapreduce: simplified data processing on large clusters. *Communications of the ACM* 51 (1): 107–113.
21. White, Tom. 2015. *Hadoop: The definitive Guide*, 4th ed. USA: OReilly Media.
22. The Apache Software Foundation. Hadoop map reduce framework. 2008. https://hadoop.apache.org/docs/r1.2.1/mapred_tutorial.html. Accessed 08 Apr 2013.
23. Grzegorz Malewicz, Matthew H Austern, Aart JC Bik, James C Dehnert, Ilan Horn, Naty Leiser, and Grzegorz Czajkowski. 2010. Pregel: A system for large-scale graph processing. In *Proceedings of the 2010 ACM SIGMOD International Conference on Management of data*, 135–146. ACM.
24. Simon Anderson. 1996. Dreamhost. https://www.dreamhost.com/. Accessed 2017.

Part II
Multi-robot Task Allocation

Chapter 6
General Background on Multi-robot Task Allocation

Abstract Multi-robot systems (MRSss) face several challenges, but the most typical problem is the multi-robot tasks allocation (MRTA). It consists in finding the efficient allocation mechanism in order to assign different tasks to the set of available robots. Toward this objective, robots will work as cooperative agents. MRTA aims at ensuring an efficient execution of tasks under consideration and thus minimizing the overall system cost. Various research works have solved the MRTA problem using the multiple traveling salesman problem (MTSP) formulation. In this context, an overview on MRTA and MTSP is given in this chapter. Furthermore, a summary of the related works is presented.

6.1 Introduction

With the progress of research in the robotics field, a wide variety of robotic application areas, such as unknown environment exploration, service robotics, the entertainment field, and so forth, have been addressed. The use of multiple robots in such complex applications has grown and thereby will ease several tasks and may lead to a higher accuracy. An MRS can improve the performance of several challenging applications that are difficult to be handled by a single robot. By collaborating together, robots are able to solve complex problems in an efficient and robust manner. The solving of a complex problem using a multi-robot system begins by dividing it into subtasks and then by dispatching these subtasks to the set of available robots. For that purpose, the overall system will be efficiently maintained and becomes more robust. Also, the use of an MRS has the advantages of reducing the whole mission completion time. It also offers a greater flexibility on the system design since robots may have different capabilities to perform all the possible tasks. The problem roughly consists of finding an optimal allocation of tasks among several robots to reduce the mission cost to a minimum. Indeed, robots are typically intended to collaborate together to achieve a given goal. In order to maintain the efficiency of the system, there is a crucial need to specify coordination policies that define how tasks are assigned to the robots. A lot of intelligent algorithms have been proposed to implement intelligence into the robots for cooperatively accomplishing their tasks. Furthermore, various researchers

A. Koubaa et al., *Robot Path Planning and Cooperation*, Studies in Computational Intelligence 772, https://doi.org/10.1007/978-3-319-77042-0_6

proposed to solve the MRTA problem using the MTSP formulation. In this chapter, a description of the MRTA problem and the MTSP is given. Furthermore, a summary of the literature related to the problems is presented.

6.2 The Multi-robot Task Allocation

In robotics, the task allocation problem addresses the issue of assigning a set of robots $R = \{r_1, ..., r_n\}$ to a set of tasks $T = \{t_1, ..., t_m\}$ in a way that optimizes a global objective function. In fact, one needs to answer the common research question: *which robot should execute which task?* [1]. In other words, *which is the best and the most efficient policy to assign the T tasks to the R robots?*
Let $\mathscr{R} = 2^R$ be the set of all robot subteams, and $\mathscr{R}^{\mathscr{T}}$ be the set of all allocations of the tasks T to the robots R. The MRTA problem can be mathematically formulated as follows [2, 3]:

Given a set of tasks T a set of robots R and a cost function for each subset of robots $r \in \mathscr{R}$ *specifying the cost of completing each subset of tasks,* $c_r : 2^T \longrightarrow \mathbb{R}^+ \cup \{\infty\}$, *find the allocation* $A^* \in \mathscr{R}^T$ *that minimizes a global objective function* $C : \mathscr{R}^T \longrightarrow \mathbb{R}^+ \cup \{\infty\}$.
In [4], Gerkey and Matarić proposed three different categories of MRTA problem:

- **First axis: Single-task robots versus multi-task robots**
  In the single-task approach, robots have the capability of executing one task at most. However, in the multi-task approach, robots have the capability of executing simultaneously multiple tasks.
- **Second axis: Single-robot tasks versus multi-robot tasks**
  Using single-robot approach, each task requires exactly one robot to be accomplished. However, in the multi-robot approach, some tasks need more than one robot to be executed.
- **Third axis: Instantaneous assignment versus time-extended assignment**
  In the instantaneous assignment approach [5, 6], the information concerning the robots, the tasks, and the environment permit only instantaneous allocation and there isn't the ability to plan for future allocations. In the time-extended assignment approach, the information allows the planning for future allocations [7, 8].

It just happens that many multi-robot task allocation problems are NP-hard problems; these problems have no exact solutions in a polynomial time; therefore, heuristics and approximate algorithms are typically used to estimate solutions.

In the multi-robot field, several research works have tackled the problem of coordination among robots. Coordination in a multi-robot system is needed when several agents are collaborating together to perform a given task or are simultaneously sharing the same resources. Coordination aims at ensuring that each robot in the MRS performs its tasks safely. Thereby, the coordination strategy should be efficient and reliable. In other words, an efficient task allocation policy needs to be set. Several research works have tackled the problem of task allocation in multi-robot systems. In the literature, there are three typical approaches devised for ensuring multi-robot coordination ([3]), namely

- **centralized approach**: it uses the global knowledge available in a central agent (e.g., control station), which produces optimal (or near-optimal) solution to the allocation problem,
- **distributed approach**: it makes decisions (or solutions) based on local information available for each agent (e.g., robot) to perform a task,
- **market-based approach**: it assumes that the tasks allocation is based on a bidding-auctioning procedure between the working agent (e.g., robot) and a coordinator that announces and allocates tasks for low-cost bidders.

The following subsections give a detailed description on these solutions.

6.2.1 Centralized Approaches

The basic idea of such approaches is the use of a single agent as a leader to ensure coordination among the entire team of robots. In fact, the leader, known also as base station, will be the unique responsible for the task allocation to the set of available robots. The base has generally global knowledge about the network status and all exchanged traffic between robots. According to [9, 10], the centralized mechanism is less robust than the distributed one. Centralized approaches can usually produce optimal plans considering the assumption of small team member and the static environment. On the other hand, these approaches suffer from a lot of points of failure such as the malfunctioning of the central agent during the mission.

6.2.2 Distributed Approaches

These approaches are more complex than the centralized ones. The decision will be made distributed in the robot team side and not commanded by a base station. Distributed systems are more robust and fault tolerant. In case of robot malfunctioning, the system will continue working; however in case of failure of the leader in the centralized approach, the whole system fails to complete the mission. In the literature, there are several distributed approaches proposed to solve the MRTA problem. For instance, Parker [11] proposed a software architecture, ALLIANCE, based on the fully distributed behavior. ALLIANCE architecture allows mobile robots to individually determine their appropriate actions to perform. Also, this architecture allows robots to handle robustly, reliably, coherently, and flexibly to abrupt environmental changes such as the addition of new mobile robots or the failure of robots. Weger and Mataric [12] proposed the broadcast of local eligibility (BLE) technique to ensure a fully distributed robots coordination in order to solve the cooperative multi-robot observation of multiple moving targets problem. BLE is based on the comparison between the robots behaviors of robots. When a robot starts the task execution, it inhibits the other robots to have the peer behavior. By that means, the robot asks for this task. In BLE architecture, re-assignment occurs almost continuously. While the cross-inhibition is an active process, in case the robot fails to execute the required task, one of the other robots will immediately take over the task.

6.2.3 Market-Based Approaches

6.2.3.1 Basics of Market-Based Approaches

The market-based approach can be stated as trade-off solution between centralized and distributed approaches. It eliminates the need for global information maintenance at the control station, while it provides more efficient solutions as compared to distributed approaches. Market-based approaches may be defined based on the following requirements [3]:

- The team is given an objective that can be decomposed into subcomponents achievable by individuals or subteams. The team has access to a limited set of resources with which to meet this objective.
- A global objective function quantifies the system designer's preferences over all possible solutions.
- An individual utility function (or cost function) specified for each robot quantifies that robot's preferences for its individual resource usage and contributions toward the team objective given its current state. Evaluating this function cannot require global or perfect information about the state of the team or team objective. Subteam preferences can also be quantified through a combination of individual utilities (or costs).
- A mapping is defined between the team objective function and individual and subteam utilities (or costs). This mapping addresses how the individual production and consumption of resources and individuals' advancement of the team objective affect the overall solution.
- Resources and individual or subteam objectives can be redistributed using a mechanism such as an auction. This mechanism accepts as input teammates bids, which are computed as a function of their utilities (or costs), and determines an outcome that maximizes the mechanism controlling agents utility (or minimizes the cost). In a well-designed mechanism, maximizing the mechanism controlling agents utility (or minimizing cost) results in improving the team objective function value.

Auctions Process

Auctions are the most important mechanism in market-based approaches. In an auction process, an auctioneer offers several items during an announcement phase (also known as bidding phase). Afterward, the participants submit bids on the offers to the auctioneer. After receiving all bids or exceeded deadline, the bidding phase finishes and a winner-determination phase begins. In this phase, the auctioneer makes decision based on the received bids from all the participants. It decides which items to award and to whom. The offered items in a robotic application may be tasks, role, or even resources. The bid prices reflect the robots' cost associated with completing a task, satisfying a role or utilizing a resource.

In the literature, several auction types can be found. The simplest one is the single-item auction [13]. In such auction, the auctioneer suggests only one item. All participants are able to submit only one bid. After receiving bids from the participants, the auctioneer gives the item to the bidder with the highest bid. In addition, there is the combinatorial auction [14, 15] in which a lot of items can be offered and each participant may bid on subsets of these items.

Cost Function

In the auction mechanism, the auctioneer allocates tasks to robots based on the cost that necessitates each one to perform each task. For that purpose, robots with the lowest costs (that is the best cost) will win the tasks. The goal is to minimize the overall cost. Estimating cost value for bids may be difficult. The cost function depends on the models of the world state and requires computationally expensive operations. For instance, the cost to reach a target position requires to have an accurate map of the environment. When the number of goal positions is important, calculating the cost to perform each task needs solving path planning problems as an instance of the traveling salesman problem that is NP-hard problem. Thus, heuristics and approximation algorithms are commonly used, implying that bid prices may not always be entirely accurate. Inaccurate bids can result in tasks not being awarded to the best robots able to complete them. In this case, re-auctioning tasks can often improve solution quality.

6.2.3.2 Previous Works

Viguria et al. proposed a distributed market-based approach called S+T (Services and Tasks) algorithm to solve the MRTA problem [16]. This latter includes the concept of services, which means that a robot provides services to another in order to execute its task. The proposed approach (S+T) is based on two roles: bidder and auctioneer. In the bidding process, if a task requires a service to be executed, it is necessary to send an initial bid to the auctioneer, which must calculate the best bid that has the lowest cost. To demonstrate the efficiency of their solution, the authors considered a surveillance mission. Using different communication ranges, they calculated the total distance traveled by all the robots, the number of messages received by one robot and the executed services. It has been shown that the communication range affects the distance traveled and the number of services (the total distance traveled by all the robots decrease when the communication range increases as far as the probability to require a service decreases). As a result, The S+T algorithm could help to complete tasks that need more than one robot to be executed.

In another work, Elmogy et al. proposed a solution for the multi-robot task allocation problem in mobile surveillance systems ([17–19]). This solution is based on the market mechanism. The authors focused on dynamic task allocation of a complex task. Complex tasks are the tasks that can be divided into subtasks. Authors in [19] defined complex task as follows: Given a set of mobile sensing agents S and a set of tasks T. Let $G \subset T$ is a group or a bundle of tasks that is decomposable into other tasks $M \in G$. The complex task allocation can be defined by a function $B : M \longrightarrow S$, mapping each subtask to a mobile sensing agent to be responsible for completing it. Equivalently, S_M is the set of all allocations of subtasks M to the team of sensors S. Their problem consists in the allocation of a set of tasks to a group of robots, in addition to the coordination between robots. To solve their problem, the authors proposed centralized and hierarchical, dynamic and fixed tree task allocation approaches. In the fixed approach, there is a set of robots used to accomplish a given task. Each one can execute one task at once, and each task can be finished by one robot. Each robot should cooperate with other members of the team to ensure a good result and to accomplish the task with a minimum cost. In the dynamic tree task allocation approach, all robots estimate their abilities to accomplish the mission task based on their plans. To validate their solution, the authors calculated the average cost as a function of the number of deployed robots. They concluded that the hierarchical dynamic tree allocation approach outperforms all the other approaches.

According to [20], the MRTA can be considered as an optimal assignment problem. This problem requires that no task is assigned to two different agents while minimizing the total cost of the assignment. In [21], the authors solved the linear assignment problem in the context of networked systems. Due to the limited communication and the lack of information exchanged between agents, a distributed auction algorithm was proposed. Every agent chooses a task and bid for it. In each iteration of the auction algorithm, each agent update the prices and highest bidders for all tasks. In that way, it is guaranteed that each agent obtains the up-to-date of both the price of the tasks and the corresponding highest bidders. The last step of each iteration is to check, for each agent, the price of the current assignment whether it has been increased by other agents in the network or whether a larger-indexed agent has placed an equal bid. Simulation results show that the auction algorithm converges to an assignment that maximizes the total assignment benefit within a linear approximation of the optimal.

In [22], the authors solve the following task assignment problem: Given n identical agents and m tasks where $m < n$ determine distributed control laws that split the agents into m groups of size n_k associated with each task $k = 1, ..., m$. A market-based solution was proposed to solve that problem. It is assumed that all agents have a prior knowledge about the tasks and maximum number of agents needed to perform each task. Neighboring groups of agents communicate together in order to compare bids and the information propagated on the network considering the availability of the requested tasks.

The initial formation problem is considered as a special case of the MRTA problem. To solve that problem, in [23], a new algorithm called the robot and task mean allocation algorithm (RTMA) was proposed. The main goal is to decide which robot should go to each of the positions of the formation while minimizing a global cost. The cost function used is equal to the difference between the distance from the robot to the task and the mean of distances from all the robots to that task. Therefore, each robot will win the task that is best for the team, not just for itself. In [24], the authors proposed a solution for the multi-robot task assignment problem with set precedence constraints (SPC-MAP). The tasks are divided into a set of groups linked by precedence constraints. It is assumed that a robot can only execute one task from each group. after finishing its allocated tasks, the robot takes benefits. The SPC-MAP is formulated as follows: Given n_r robots, n_t tasks with the tasks divided into ns disjoint subsets, maximize the total benefits of robot task assignment with the set precedence constraints for tasks, such that, each task is performed by one robot, and each robot r_i performs exactly N_i tasks and at most one task from each subset. To solve this problem, an auction-based algorithm has been proposed. In the beginning of each iteration, each robot must update its assignment information from the previous iteration. To bid for a task, the following constraints should be satisfied: (a) robot r_i is assigned to exactly N_i tasks; (b) robot r_i is assigned to at most one task in each subset.

In [25], the authors solved the multi-robot task assignment problem with task deadline constraints (DiMAP) which is an extension of the SPC-MAP problem. Considering the assumption of limited battery life, each robot can execute a limited number of tasks. The goal is to find the allocation for the tasks to the robots with the minimum cost while achieving the deadline constraints. The problem is formulated as follows: given a set of tasks T, with each task $t_j \in T$ having a deadline d_j. Each task has to be done by only one robot, and each robot can do one task at a time. The maximum number of tasks that robot r_i can do is N_i . Each robot r_i obtains a payoff a_{ij} for doing task t_j. The overall payoff is the sum of the individual robot payoffs. To solve the DiMAP problem, the authors proposed an auction-based distributed algorithm. At each iteration, a robot r_i receives the local task price from its neighbors. It updates its own price and calculates its task value. Next, the robot r_i selects a task set according to the following conditions:*(a)* robot r_i is assigned to at most N_i tasks; *(b)* r_i is assigned to at most k tasks of all tasks with deadline no more than k. Finally, the robot r_i is assigned to task set, and it updates the task price. At the end of the algorithm, each robot is assigned to its task. The algorithm was analyzed in terms of soundness (does the final solution satisfy all constraints?), completeness (Will the solution be found in a finite number of iterations and be feasible?) and optimality (How good is the solution?). The performance evaluation of a scenario of 20 robots, 100 tasks, and 5 tasks per robots shows that there is a trade-off between the solution quality and the convergence time.

In [26], the authors addressed the multi-robot patrolling task. The authors designed an approach for monitoring and reassigning tasks. In the beginning, the patrolling area is divided into partitions. Every partition is allocated to a robot which must visit all the nodes belonging to its partition. When the refresh time of the assigned nodes of a robot reaches a threshold, a central monitor process uses an auction-based protocol in order to perform a task re-assignment. After receiving an announcement for a candidate node, each robot adds this node to its patrol partition and calculates the new bid. The robot that has the best refresh time (i.e., the minimum value) will be assigned to the node. To evaluate the performance of the designed approach, three different scenarios (3, 5, and 8 robots) were run for 2 hours. In the experimental study, the authors used a team of three TurtleBots scattered in the same office environment. The comparison to a naive approach, which does not consider individual robot performance, shows that the results obtained by the auction strategy improve the refresh time.

Paper [27] addressed the problem of assigning n robots to execute n tasks. The solution proposed is a distributed version of the well-known Hungarian method [28]. The goal is to find finding the assignment with the minimum total cost (distance) between the robots and the tasks. Each robot is assumed to make decision based on the information it has, such as the distance to the target, and the information received from the other robots. The communication between robots is performed over a connected dynamic communication network. It is shown that the computational time performed by each robot (in the order of $O(n^2)$) results minor than the time required by the standard (centralized) Hungarian algorithm.

6.3 The Multiple Traveling Salesman Problem

6.3.1 MTSP Overview

The traveling salesman problem (TSP) is one of the classic, well-known problems of operations research. The TSP is the problem of planning a route of a salesman ($n = 1$) to visit a set of m nodes, with the condition of visiting each node only once, starting from and ending at a node called node depot. The objective is to find the best route, that is the route with the minimum distance. TSP belongs to the NP-hard problems, because the time increases exponentially with the increase in the number of cities. The MTSP is the generalized form of the TSP, where more than one salesman ($n > 1$), sharing the same workspace, are involved in visiting a set of $m > n$ cities such that each city must be visited exactly once with the objective of minimizing the total cost of visiting all cities. Accordingly, two issues must be solved: (1) how to separate the m cities into n groups and (2) how to order the cities for each route to minimize the distance.

Depending on the number of depots d, the MTSP can be classified into several versions:

- $d = 1$: All routes start from and end at the same depot d.
- $d = n$: Each salesman starts from and ends at its own depot while all depots are different from each other. Therefore, in this case, the tour of each salesman is closed.
- $d = 2 * n$: In this case, the tour of each salesman is open. Every salesman has two depots, one as a start point and the second as an end point.
- $1 < d < 2m$: This case combines the above three cases. In fact, more than one salesman can share the same depot.

The MTSP is defined on a graph $G = (V, A)$ [29], where V is the set of cities and A is the set of arcs between the cities. Let c_{ij} be the cost associated with the arc (i, j) and x_{ij} the binary variable that is equal to 1 if the arc (i, j) belongs to the condidate tour (i.e. the tour of the current salesman) and 0 otherwise.

6.3.2 Related Works on MTSP

6.3.2.1 Genetic Algorithm-Based Solutions

To solve NP-hard combinatorial optimization problems, such as the MTSP, heuristics, like GAs [30], are widely used [31–34].
To solve the MTSP, Yuan et. al [35] used the two-part chromosome representation and proposed a new crossover operator. The new solution was compared to other crossover methods, including the ordered crossover operator (ORX), the cycle crossover operator (CYX), and the partially matched crossover operator (PMX). The authors used two different objective functions, namely the total traveled distance by all salesmen and the longest route among all the salesmen. From the experiments, it was proved that the new crossover method enables the genetic algorithm to produce higher solution quality.

Grouping genetic algorithms (GGA) [36] are a variation of the GA. In [37], a GGA was developed for the MTSP. The concept is to group the cities into routes, where the number of routes is the same as the number of salesmen and then order the cities on each route. For the comparison, the authors considered two objectives, the maximum tour length and the total tour length. From the simulation study, it is clearly proved that the GGA produces better solutions when the objective is to minimize the maximum tour length as the GGA is designed for grouping problems. However, the performance of the GGA, when the objective is to minimize the total tour length, degrades.
Moreover, another GGA called steady-state grouping genetic algorithm (GGA-SS) was proposed in [38]. In this paper, the authors used a different chromosome representation and genetic operators from those used in [37]. Similar to several works, stated in the literature, the aim is to minimize the total distance traveled by all the salespersons and the distance traveled by any salesperson. The authors used the test problems proposed in [37, 39]. The given results show that the GGA-SS is able to produce better results in terms of both the objectives in comparison with the solutions found in [37, 39].

6.3.2.2 Ant Colony Optimization-Based Solutions

The ACO algorithm is an iterative search technique inspired from the behavior of real ant colonies. In the literature, ACO algorithm was used to solve several robotic problems such as path planning [40, 41], coordination [42]. In [43], an ACO algo-

rithm was used to solve the MTSP with the objective of minimizing the maximum tour length of all the salesmen and the maximum tour length of each salesman. The simulation results of the algorithm show that the ACO algorithm outperforms the GA-based algorithms proposed in [37–39].
Also, a modified ACO algorithm (MACO) was proposed in [44]. To improve the quality of the solution, the authors suggest modifications including the transition rule, the candidate list, the global pheromone updating rules, and several local search techniques. The aim of the algorithm is to minimize the distance traveled by the salesmen. The solution was tested on standard benchmarks available from the literature. Computational results have shown that the MACO algorithm gives better solutions in comparison with the existing solutions for the MTSP.

6.3.2.3 Other Heuristic Solutions

In [45], the authors studied the MTSP and proposed two metaheuristic approaches. The first approach is based on the artificial bee colony (ABC) algorithm [46], whereas the second approach is based on invasive weed optimization (IWO) algorithm [47]. Both approaches were evaluated using the benchmark instances existing in the literature. The comparison results with the state-of-the-art solutions in terms of total distance traveled by all the salespersons, and the maximum distance traveled by any salesperson demonstrate that the ABC and IWO algorithms give better results for both objectives.

6.3.2.4 Market-Based Solutions

Other solutions called market-based solutions were used to solve the MTSP.

In [48], the authors proposed a market-based approach to solve the multiple depot MTSP. The algorithm consists of four steps: market auction, agent-to-agent trade, agent switch, and agent relinquish step. Three performance criteria were considered: the quality of the solution, the number of iterations required to get a solution, and the execution time. It was shown that the solution generates better solutions in comparison with other suboptimal solutions.

As several real-world applications need to optimize multiple objectives, we addressed the multi-objective problem and we propose a solution that optimizes several objectives simultaneously, including the total traveled distance, the maximum tour length, etc.

6.3.3 Multi-objective Optimization Problem (MOP)

6.3.3.1 MOP Overview

A multi-objective optimization problem can be formulated through a mathematical model defined by a set of p objective functions, which must be minimized or maximized. Note that the objectives are usually in conflict with one another. Thus, it is impossible to simultaneously improve all objective functions. This means that there is no single efficient solution; instead, there is more than one optimal solution that provides a balanced optimization of the different metrics, also known as Pareto-optimal solutions [49]. A general multi-objective optimization problem can be defined as follows:

$$\begin{aligned} min/max \quad & f1(X) \\ min/max \quad & f2(X) \\ & . \\ & . \\ & . \\ min/max \quad & fp(X) \end{aligned} \tag{6.1}$$

where $f_1(X), f_2(X), ..., f_p(X)$ are the p $(>= 2)$ conflicting objective functions, and $X = (x_1, x_2, ..., x_n)^T$ belongs to the feasible region $S \subset R^n$.

6.3.3.2 Related Works on MOP

In the literature, few researchers considered the MTSP as a multi-objective optimization problem.
In [50], the authors proposed a multi-objective, non-dominated sorting genetic algorithm (NSGA-II) to treat the MTSP. The objectives to be optimized are the total traveled distance and the working times by the salesmen. The authors are intended to find a set of non-dominated solutions that, when compared, some are better in one/some objective(s), and others in other objectives. To evaluate the performance of the proposed solution, two instances ((3 salesmen, 29 nodes) and (3 salesmen, 75 customers)) were considered. The results show the effectiveness of the NSGA-II in minimizing both objectives.

In [51], the authors used the ACO algorithm for solving the task assignment problem for multiple unmanned underwater vehicles. The authors aim to optimize two objectives, namely the total distance of visiting all targets and the total turning angle while considering the constraint of balancing the number of targets visited by each vehicle. The solution consists of two phases. The first phase is the task number assignment phase. It consists of defining the number of targets for each vehicle.

The second phase solves the MTSP problem using an ant colony for each objective. Performance evaluation shows that the algorithm generates good solution.

The market-based approach is widely used to solve several problems including the MTSP. In [52], the solution consists of using a clustering technique with an auction process. The objectives are to minimize the distance traveled by all the robots and balance the workload between the robots equally. The first step is to decompose N tasks into n groups in such a way the distance inside each cluster is minimized. Then, the cost for each robot to visit the n clusters is computed. Finally, in the auction step, each cluster is allocated to the robot which provides the lowest bid. We noticed that the complexity of the algorithm is relatively high because all possible combinations of the assignment of clusters to robots are considered. This means that the complexity of the algorithm increases with the number of clusters. To evaluate the performance of the algorithm, the authors used the benchmark VRP dataset "A-n32-K5.vrp" [53]. The total cost used to assign a cluster is equal to the sum of the cost of visiting the tasks in the cluster and the idle cost (i.e., sum of the difference in cost of travel between any two robots). Two scenarios were considered: one with two clusters and the other with three clusters. We noticed that the used scenarios are not sufficient to prove the effectiveness of the algorithm.
Moreover, in [54], an auction algorithm using a clustering technique has been proposed with the objectives of minimizing both the maximum traveled distance of each robot and the sum of distance traveled by all robots in visiting their assigned locations. The algorithm process is as follows. Initially, it is assumed that all robots have a list of allocated tasks. In the case where a robot reaches the position of a task, it sends a specific signal to the other robots to be able to make auctions. After all auctions are completed, the robot re-plans its path and move to the next task. If a robot receives a message to make an auction, it forms a new set of clusters of its assigned tasks and then an auction process starts for the newly formed clusters except the cluster that contains its currently initialized task. When a robot receives an auction for a cluster, it bids for that cluster. Finally, the robot with the lowest bid will win the cluster. For the performance evaluation, the authors only have shown the percentage of improvement of the initial assignment as compared to the final assignment.

In [55], the authors proposed a new method for multi-objective traveling salesman called two-phase Pareto local search (2PPLS). In the first phase, each single-objective problem is solved separately using one of the best heuristics for the single objective. In the second phase, a Pareto local search is applied to every solution of the initial phase using a 2-opt neighborhood with candidate lists. It is important to note that there is a need to solve a high number of weighted single-objective problems, before applying the Pareto local search, which may cause efficiency degradation. Also, the integration of the 2-opt process may achieve poor effectiveness with low efficiency when the number of feasible objective vectors is small, whereas it obtains desired effectiveness with low efficiency when the number of feasible objective vectors is large.

In [56], authors integrated ant colony optimization (ACO) to local search technique in order to solve the multi-objective Knapsack problems (MOKPs) and the multi-objective traveling salesman problem (MTSPs). MOEA/D-ACO decomposes a multi-objective optimization problem into various single objective optimization problems. Each ant is assigned to a subproblem, and each ant has several neighboring ants. A heuristic information matrix is maintained by each ant. The main issue related to this approach is the uncertainty of the time convergence and the implementation complexity. Shim et al. presented in [57] a mathematical formulation of the multi-objective multiple traveling salesman problems (MOmTSP). The proposed approach is not required to differentiate between the dominated and non-dominated solutions. Its objective is to determine cycles and cover a set of potential customers in order to maximize the corresponding benefit and minimizing the total traveled distance. An estimation of distribution algorithm (EDA) with a gradient search is used for the solution of the considered problem. The proposed approach works well when the users have prior knowledge about the problem to assign weights.

In [58], the authors presented a detailed comparison between MOEA/D and NSGA-II. The paper focuses on the performance of the multi-objective traveling salesman problem and studies the effect of local search on the performance of MOEA/D. Compared to MOEA/D, NSGA-II has no bias in searching any particular part of the Pareto front. All non-dominated solutions in the current population have an equal chance to be selected for reproduction. However, this might not be efficient when sampling offspring solutions due to the following reasons. First, the non-dominated solutions might have very different structures in the decision space. Therefore, the possibility of generating high-quality offspring solutions by recombining these solutions is low. Second, the design of recombination operators is often problem dependent. In MOEA/D, weight vectors and aggregate functions play a very important role to solve various kinds of problems. Overall MOEA/D has shown much better algorithmic improvement than NSGA-II.

6.4 Conclusion

This chapter provides a background on the MRTA problem and MTSP. A brief overview of research on MRTA problem and MTSP is provided in this chapter. These problems are well known in robotics, and various research works have been proposed to solve them. A brief summary of the concept of market-based approach was described. A summary of the main research works related to market-based algorithms to solve the MRTA and the MTSP is explained. The research works on MTSP have been classified into two categories: single objective algorithms and multi-objective algorithms. Heuristics like GA and ACO are widely used to solve this problem. Several heuristic-based algorithms have briefly been reviewed here. In the next chapter, we will propose different approaches investigating these problems.

References

1. Diego, Pizzocaro and A. Preece. 2009. Towards a taxonomy of task allocation in sensor networks. In *Proceedings of the 28th IEEE international conference on Computer Communications Workshops (INFOCOM)*, 413–414.
2. Nidhi, Kalra, Robert Zlot, M. Bernardine Dias, and Anthony Stentz. 2005. Market-based multirobot coordination: A comprehensive survey and analysis. Technical report, CARNEGIE-MELLON UNIV PITTSBURGH PA ROBOTICS INST.
3. Dias Bernardine, M., Robert Zlot, Nidhi Kalra, and Anthony Stentz. 2006. Market-based multirobot coordination: A survey and analysis. *Proceedings of the IEEE* 94(7):1257–1270.
4. Brian, P.G. and M.J. Matarić. 2004. A formal analysis and taxonomy of task allocation in multi-robot systems. *The International Journal of Robotics Research* 23(9):939–954.
5. Sylvia C. Botelho and Rachid Alami. 1999. M+: a scheme for multi-robot cooperation through negotiated task allocation and achievement. In *IEEE International Conference on Robotics and Automation, Proceedings*, vol. 2, 1234–1239. IEEE.
6. Gerkey, B.P. and M.J. Mataric. 2002. Sold!: Auction methods for multirobot coordination. *IEEE transactions on robotics and automation* 18(5):758–768.
7. Fang, Tang and Spondon Saha. 2008. An anytime winner determination algorithm for time-extended multi-robot task allocation. In *ARCS*, 123–130.
8. Zlot, Robert, and Anthony Stentz. 2006. Market-based multirobot coordination for complex tasks. *The International Journal of Robotics Research* 25(1):73–101.
9. Barry L, Brumitt and Anthony Stentz. 1998. Grammps: A generalized mission planner for multiple mobile robots in unstructured environments. In *IEEE International Conference on Robotics and Automation, Proceedings.*, vol. 2, 1564–1571. IEEE.
10. Philippe, Caloud, Wonyun Choi, J.-C. Latombe, Claude Le Pape, and Mark Yim. 1990. Indoor automation with many mobile robots. In *IEEE International Workshop on Intelligent Robots and Systems' 90. 'Towards a New Frontier of Applications', Proceedings. IROS'90.*, 67–72. IEEE.
11. Parker, Lyne E. 1998. Alliance: An architecture for fault tolerant multirobot cooperation. *IEEE transactions on robotics and automation* 14 (2): 220–240.
12. Barry Brian, Werger and M.J. Matarić. 2000. Broadcast of local eligibility for multi-target observation. In *Distributed autonomous robotic systems 4*, pages 347–356. Springer, Berlin.
13. Sanem, Sariel and Tucker R. Balch. 2006. Efficient bids on task allocation for multi-robot exploration. In *FLAIRS Conference*, 116–121.
14. Sven Koenig, Craig A. Tovey, Xiaoming Zheng, and Ilgaz Sungur. 2007. Sequential bundle-bid single-sale auction algorithms for decentralized control. In *IJCAI*, 1359–1365.
15. Liu, Lin and Zhiqiang Zheng. 2005. Combinatorial bids based multi-robot task allocation method. In *Proceedings of the 2005 IEEE International Conference on Robotics and Automation, ICRA*, 1145–1150. IEEE.
16. Antidio, Viguria, Ivan Maza, and Anibal Ollero. 2008. S+ t: An algorithm for distributed multi-robot task allocation based on services for improving robot cooperation. In *IEEE International Conference on Robotics and Automation ICRA*, 3163–3168. IEEE.
17. Ahmed M. Elmogy, Alaa M. Khamis, and Fakhri O. Karray. 2009. Dynamic complex task allocation in multisensor surveillance systems. In *2009 3rd International Conference on Signals, Circuits and Systems (SCS)*, 1–6. IEEE.
18. Ahmed M. Elmogy, Alaa M. Khamis, and Fakhri O. Karray. 2009. Market-based dynamic task allocation in mobile surveillance systems. In *2009 IEEE International Workshop on Safety, Security and Rescue Robotics (SSRR)*, 1–6. IEEE.
19. Alaa M. Khamis, Ahmed M. Elmogy, and Fakhri O. Karray. 2011. Complex task allocation in mobile surveillance systems. *Journal of Intelligent and Robotic Systems* 64(1):33–55.
20. Yan, Zhi, Nicolas Jouandeau, and Arab Ali Cherif. 2013. A survey and analysis of multi-robot coordination. *International Journal of Advanced Robotic Systems* 10(12):399.

21. Michael M. Zavlanos, Leonid Spesivtsev, and George J. Pappas. 2008. A distributed auction algorithm for the assignment problem. In *47th IEEE Conference on Decision and Control CDC*, 1212–1217. IEEE.
22. Nathan, Michael, Michael M. Zavlanos, Vijay Kumar, and George J. Pappas. 2008. Distributed multi-robot task assignment and formation control. In *IEEE International Conference on Robotics and Automation ICRA*, 128–133. IEEE.
23. Antidio, Viguria and Ayanna M. Howard. 2009. An integrated approach for achieving multi-robot task formations. *IEEE/ASME Transactions on Mechatronics* 14(2):176–186.
24. Lingzhi, Luo, Nilanjan Chakraborty, and Katia Sycara. 2011. Multi-robot assignment algorithm for tasks with set precedence constraints. In *2011 IEEE International Conference on Robotics and Automation (ICRA)*, 2526–2533. IEEE.
25. Lingzhi, Luo, Nilanjan Chakraborty, and Katia Sycara. 2013. Distributed algorithm design for multi-robot task assignment with deadlines for tasks. In *2013 IEEE International Conference on Robotics and Automation (ICRA)*, 3007–3013. IEEE.
26. Charles, Pippin, Henrik Christensen, and Lora Weiss. 2013. Performance based task assignment in multi-robot patrolling. In *Proceedings of the 28th annual ACM symposium on applied computing*, 70–76. ACM.
27. Stefano, Giordani, Marin Lujak, and Francesco Martinelli. 2010. A distributed algorithm for the multi-robot task allocation problem. In *International Conference on Industrial, Engineering and Other Applications of Applied Intelligent Systems*, 721–730. Springer, Berlin.
28. Harold W. Kuhn. 1955. The hungarian method for the assignment problem. *Naval Research Logistics (NRL)*, 2(1–2):83–97.
29. Bektas, Tolga. 2006. The multiple traveling salesman problem: an overview of formulations and solution procedures. *Omega* 34(3):209–219.
30. John H. Holland. 1975. *Adaptation in natural and artificial systems: an introductory analysis with applications to biology, control, and artificial intelligence.* U Michigan Press, Michigan.
31. András, Király and János Abonyi. 2011. Optimization of multiple traveling salesmen problem by a novel representation based genetic algorithm. In *Intelligent Computational Optimization in Engineering*, 241–269. Springer, Berlin.
32. Jun, Li, Qirui Sun, MengChu Zhou, and Xianzhong Dai. 2013. A new multiple traveling salesman problem and its genetic algorithm-based solution. In *2013 IEEE International Conference on Systems, Man, and Cybernetics (SMC)*, 627–632. IEEE.
33. Harpreet, Singh and Ravreet Kaur. 2013. Resolving multiple traveling salesman problem using genetic algorithms. *International Journal of Computer Science Engineering* 3(2):209–212.
34. Shalini, Singh and Ejaz Aslam Lodhi. 2014. Comparison study of multiple traveling salesmen problem using genetic algorithm. *International Journal of Computer Science and Network Security (IJCSNS)* 14(7):107–110.
35. Yuan, Shuai, Bradley Skinner, Shoudong Huang, and Dikai Liu. 2013. A new crossover approach for solving the multiple travelling salesmen problem using genetic algorithms. *European Journal of Operational Research* 228(1):72–82.
36. Falkenauer, Emanuel. 1992. The grouping genetic algorithms-widening the scope of the gas. *Belgian Journal of Operations Research, Statistics and Computer Science* 33(1):79–102.
37. Evelyn, C. 2007. Brown, Cliff T Ragsdale, and Arthur E Carter. A grouping genetic algorithm for the multiple traveling salesperson problem. *International Journal of Information Technology and Decision Making* 6(02):333–347.
38. Alok, Singh and Anurag Singh Baghel. 2009. A new grouping genetic algorithm approach to the multiple traveling salesperson problem. *Soft Computing-A Fusion of Foundations, Methodologies and Applications* 13(1):95–101.
39. Arthur E. 2006. Carter and Cliff T Ragsdale. A new approach to solving the multiple traveling salesperson problem using genetic algorithms. *European Journal of Operational Research* 175(1):246–257.
40. Imen Châari, Anis Koubaa, Hachemi Bennaceur, Sahar Trigui, and Khaled Al-Shalfan. 2012. Smartpath: A hybrid aco-ga algorithm for robot path planning. In *2012 IEEE Congress on Evolutionary Computation (CEC)*, 1–8. IEEE.

41. Châari, Imen, Anis Koubâa, Sahar Trigui, Hachemi Bennaceur, Adel Ammar, and Khaled Al-Shalfan. 2014. Smartpath: An efficient hybrid aco-ga algorithm for solving the global path planning problem of mobile robots. *International Journal of Advanced Robotic Systems* 11(7):94.
42. Anis, Koubâa, Sahar Trigui, and Imen Châari. 2012. Indoor surveillance application using wireless robots and sensor networks: Coordination and path planning. *Mobile Ad Hoc Robots and Wireless Robotic Systems: Design and Implementation*, 19–57.
43. Weimin, Liu, Sujian Li, Fanggeng Zhao, and Aiyun Zheng. 2009. An ant colony optimization algorithm for the multiple traveling salesmen problem. In *4th IEEE Conference on Industrial Electronics and Applications ICIEA*, 1533–1537. IEEE.
44. Yousefikhoshbakht, Majid, Farzad Didehvar, and Farhad Rahmati. 2013. Modification of the ant colony optimization for solving the multiple traveling salesman problem. *Romanian Academy Section for Information Science and Technology* 16(1):65–80.
45. Venkatesh, Pandiri, and Alok Singh. 2015. Two metaheuristic approaches for the multiple traveling salesperson problem. *Applied Soft Computing* 26:74–89.
46. Dervis, Karaboga. 2005. An idea based on honey bee swarm for numerical optimization. Technical report, Technical report-tr06, Erciyes University, Engineering Faculty, Computer Engineering Department.
47. Ali Reza, Mehrabian and Caro Lucas. 2006. A novel numerical optimization algorithm inspired from weed colonization. *Ecological Informatics* 1(4):355–366.
48. Elad, Kivelevitch, Kelly Cohen, and Manish Kumar. 2013. A market-based solution to the multiple traveling salesmen problem. *Journal of Intelligent and Robotic Systems*, 1–20.
49. Kaisa, Miettinen. 2012. *Nonlinear multiobjective optimization*, vol. 12. Springer Science and Business Media, Berlin.
50. Bolaños, R., M. Echeverry, and J. Escobar. 2015. A multiobjective non-dominated sorting genetic algorithm (nsga-ii) for the multiple traveling salesman problem. *Decision Science Letters* 4(4):559–568.
51. Zhenzhen, Xu, Yiping Li, and Xisheng Feng. 2008. Constrained multi-objective task assignment for uuvs using multiple ant colonies system. In *ISECS International Colloquium on Computing, Communication, Control, and Management CCCM'08*, vol. 1, 462–466. IEEE.
52. Elango, Murugappan, Subramanian Nachiappan, and Manoj Kumar Tiwari. 2011. Balancing task allocation in multi-robot systems using k-means clustering and auction based mechanisms. *Expert Systems with Applications* 38(6):6486–6491.
53. TSPLIB95. http://www.iwr.uni-heidelberg.de/groups/comopt/software/TSPLIB95/.
54. Bradford Gregory John, Heap and Maurice Pagnucco. 2012. Repeated sequential auctions with dynamic task clusters. In *AAAI*, 19972002.
55. Thibaut, Lust and Jacques Teghem. 2010. *The multiobjective traveling salesman problem: A survey and a new approach*, 119–141. Springer, Berlin.
56. Ke, Liangjun, Qingfu Zhang, and Roberto Battiti. 2013. Moea/d-aco: A multiobjective evolutionary algorithm using decomposition and antcolony. *IEEE Transactions on Cybernetics* 43(6):1845–1859.
57. Vui Ann, Shim, Kay Chen Tan, and Kok Kiong Tan. 2012. A hybrid estimation of distribution algorithm for solving the multi-objective multiple traveling salesman problem. In *2012 IEEE Congress on Evolutionary Computation (CEC)*, 1–8. IEEE.
58. Wei, Peng, Qingfu Zhang, and Hui Li. 2009. Comparison between moea/d and nsga-ii on the multi-objective travelling salesman problem. *Multi-objective memetic algorithms*, 309–324.

Chapter 7
Different Approaches to Solve the MRTA Problem

Abstract The multi-robot task allocation problem is a fundamental problem in robotics research area. The problem roughly consists of finding an optimal allocation of tasks among several robots to reduce the mission cost to a minimum. As mentioned in Chap. 6, extensive research has been conducted in the area for answering the following question: *Which robot should execute which task?* In this chapter, we design different solutions to solve the MRTA problem. We propose four different approaches: an improved distributed market-based approach (IDMB), a clustering market-based approach (CM-MTSP), a fuzzy logic-based approach (FL-MTSP), and Move-and-Improve approach. These approaches must define how tasks are assigned to the robots. The IDBM, CM-MTSP, and Move-and-Improve approaches are based on the use of an auction process where bids are used to evaluate the assignment. The FL-MTSP is based on the use of the fuzzy logic algebra to combine objectives to be optimized.

7.1 Introduction

In complex robotics applications, a cooperative robot system represents a recommended alternative to a single-robot system considering the collaborative effect between robots, which leads to accomplishing their mission more efficiently. Multi-robot coordination has been a major challenge in robotics with a vast array of applications. This is the case for a disaster management scenario, in which the task is to collect sensor data and take live images from locations impacted by the disaster, to help rescuers take appropriate actions in real time and for an emergency response application, where a team of robots has to visit affected locations to provide rescue services.

This chapter describes the following approaches proposed to solve the MRTA problem.

1. An improved distributed market-based assignment algorithm [1] has been introduced. Robots are able to bid for tasks with the assumption that each robot can perform only one task and each task is assigned to only one robot at a time. An

A. Koubaa et al., *Robot Path Planning and Cooperation*, Studies in Computational Intelligence 772, https://doi.org/10.1007/978-3-319-77042-0_7

improvement phase that consists of swapping tasks is applied after finishing the assignment of the robots to the targets in order to improve the cost.
2. A new clustering market-based approach is proposed [2]. First, the target locations are divided into groups and then assigned to robots using a market-based approach. We formulated the problem as multiple depots multiple traveling salesman problem and addressed the multi-objective optimization of three objectives, namely the total traveled distance, the maximum traveled distance, and the mission time.
3. A complete new approach, based on fuzzy logic algebra [3], is used to solve the multiple traveling salesman problem while optimizing multiple performance criteria, namely the total traveled distance of all robots on the targets and the maximum traveled distance of any robot. The idea consists in modeling each criterion using a membership function that will be used as input to a fuzzification system to produce an output value representing the combined metrics. We design a centralized approach to efficiently assign target locations to robots.
4. A new market-based approach to solve the multiple depots MTSP (MD-MTSP) called *Move-and-Improve* is used. The Move-and-Improve mechanism is based on improving the solution, while robots are moving, through communication with each other. There is no prior centralized assignment of locations to the robots, but every robot collects to its best target locations by runtime negotiation with its peers until the system converges to non-overlapping target allocation to the robots.

7.2 Objective Functions

In this chapter, three are considered, namely the total tour length, the maximum tour length, and the mission time.

1. The total traveled distance of the robots on the target: We define TTD as the sum of all tours length performed by all the robots. The tour length of each robot is calculated using existing TSP solver once the target locations are allocated. The total tour length is calculated by summing up the traveled distance of all edges included in a tour. We define $tour_{r_i}$ (Eq. 7.2) as the tour of the robot i starting from and ending at the same position. The TTD is given according to Eq. 7.1.

$$TTD = \sum_{i=1}^{n} tour_{r_i} \tag{7.1}$$

where

$$\begin{aligned} tour_{r_i} &= distance(r_i, t_{i_1}) \\ &+ \textstyle\sum_{j=1}^{k_i-1} distance(t_{i_j}, t_{i_{j+1}}) \\ &+ distance(t_{i_{k_i}}, r_i) \end{aligned} \tag{7.2}$$

where k_i is the number of target locations assigned to robot i. $distance(t_{i_j}, t_{i_{j+1}})$ represents the distance between target location j and target location $j + 1$ for robot i. t_{i_1} and $t_{i_{k_i}}$ represent the first and the last target locations, respectively, for robot i. $distance(r_i, t_{i_1})$ represents the distance to travel from the depot of r_i to the first target t_{i_1}, and $distance(t_{i_{k_i}}, r_i)$ represents the distance to return back from the last target $t_{i_{k_i}}$ to the initial depot of r_i.

2. The maximum tour MT measured in terms of distance: It is the maximum traveled distance (MTD) by any robot after the scheduled mission is completed. The maximum tour length among all the tours of the robots is expressed as follows:

$$\begin{aligned} MT &= max(tour_{r_i}) \\ &1 \leq i \leq n \\ s.t.\ & tour_{r_i} \neq tour_{r_j} \\ &1 \leq j \leq n,\ i \neq j \end{aligned} \tag{7.3}$$

3. The mission time of a robot: It is the time necessary for the tour completion of a robot r_i.

$$Tr_i = tour_{r_i}/Sr_i \tag{7.4}$$

$tour_{r_i}$ represents the tour length of robot r_i as expressed in Eq. 7.2, and Sr_i represents the speed of robot r_i.

7.3 Improved Distributed Market-Based Approach

In this section, we considered scenarios where the number of targets is equal to the number of robots. The mission is to visit target positions using a group of autonomous mobile robots. The problem can be considered as a job assignment problem [4] where robots are the workers and tasks are the jobs to be executed by those workers. Although this task assignment problem was solved optimally in a centralized manner, using, for example, the Hungarian method [4], these solutions have the drawbacks of centralized systems such as the slow response to dynamic changes. The solution to the multi-robot assignment problem is based on a market process [5]. We proposed an improved distributed market-based assignment algorithm where robots are able to bid for tasks with the assumption that each robot can perform only one task and each task is assigned to only one robot at a time. The market-based approach provides a good trade-off between centralized and distributed solutions. It eliminates the need for global information maintenance at a control station, while it provides more efficient solutions than other distributed approaches where only local information is provided. Also, a market-based approach can work even without knowing exactly the number of available robots in communication range, and this number could change dynamically without compromising the operation of the whole system. Like any market-based approach, we define two main entities: the agents (i.e., robots) and the tasks. Each agent can play two roles: bidder (buyer) and auctioneer (seller). The auctioneer is the agent that announces offers of several items during an announcement phase. In our case, the items are the tasks. Then the bidders submit bids to the offers of the auctioneer.

7.3.1 Distributed Market-Based (DMB) Algorithm

The main idea of the algorithm is that each robot should have only one assigned task. Therefore, it will retain the task with the best cost (i.e., lowest distance). In order to ensure this condition, the DMB algorithm uses task reallocation process that will ameliorate the assignment. Furthermore, the number of interchanged messages between robots will increase. Since the number of tasks is equal to the number of robots, we initially assign each robot to a task. Each robot agent plays the role of the auctioneer in order to announce its task, and the rest of robots team are the bidders. The auction process starts with an announcement phase, where each robot sends an announcement message to provide its bid on the announced task. It should be pointed out that all tasks will be announced at the same time. Thus, the algorithm always produces the same solution for the same scenario. Assuming that robots are in the same communication range, all announcement messages will be received by all robots. Possible collisions can be overcome by using contention resolution protocols [6]. Each robot receives $(n - 1)$ (i.e., n is the total number of robots or tasks) announcement messages at a time, so each robot bids only for the tasks in which it has a cost lower than that it was announced.

After receiving all bids, the auctioneer allocates the task to the robot with the best bid (lowest cost). After the assignment, a robot may win more than one task. In this case, it keeps the task with the best bid and sells the others. In the reallocation process, only the robots which did not win tasks can bid for the new announced tasks. This condition prevents from having an infinite loop in the auction process. Figure 7.1 presents a scenario showing the difference between the case using the condition and the case without the condition. The reallocation process finishes when all tasks are allocated. At the end, each task will be assigned to a robot. The auctioneer and bidder algorithms for a robot are presented in Algorithms 8 and 9, respectively.

Algorithm 8: Auctioneer algorithm

```
1 if The list of tasks to announce is not empty then
2 |   Announce task
3 |   Receive bids
4 |   Select the best bid (lowest cost)
5 |   Send the task to the robot with the best bid
6 end
```

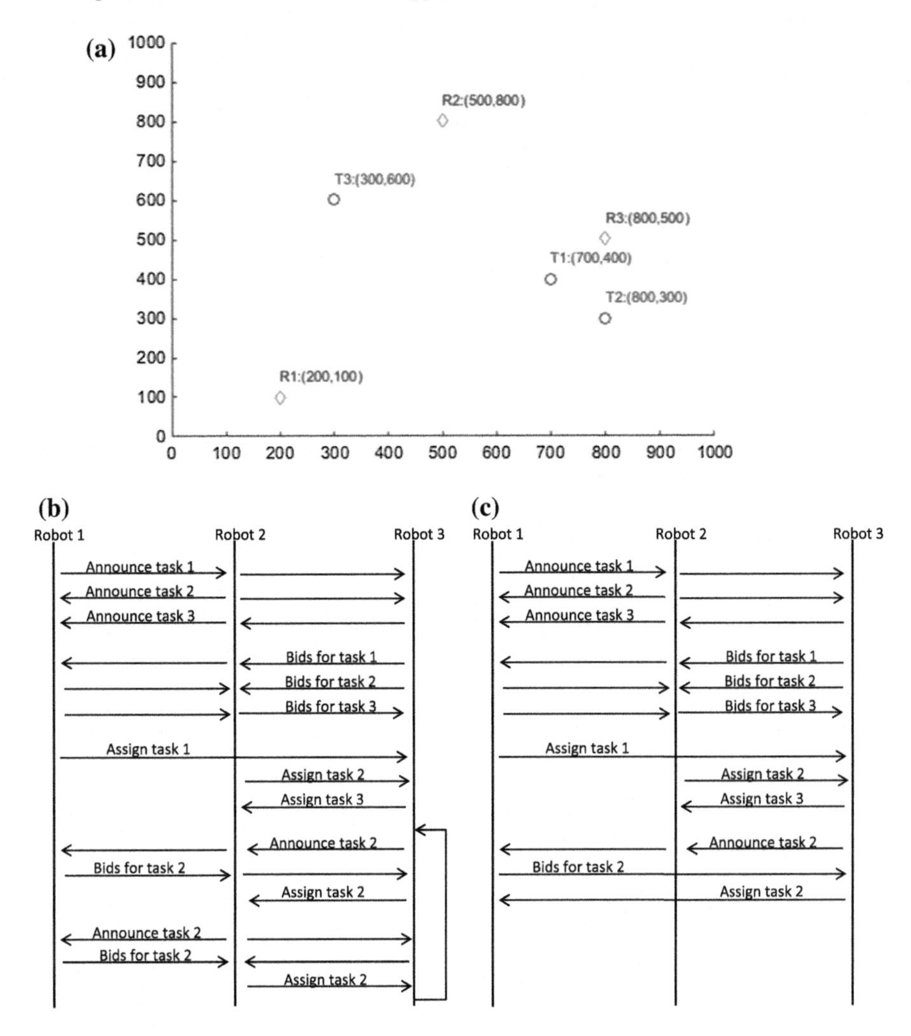

Fig. 7.1 **a** Initial position of the robots and the targets to be allocated. **b** Messages interchanged between the robots with the appearance of an infinite loop. **c** Messages interchanged between the robots for the DMB algorithm

Algorithm 9: Bidder algorithm

```
1  Receive message
2  if received message is a task announcement then
3  |   if the robot has not yet won a task then
4  |   |   Calculate bid (cost)
5  |   |   if The list of tasks to announce is not empty then
6  |   |   |   if cost of new task < cost of announced task then
7  |   |   |   |   Send bid to the auctioneer
8  |   |   |   end
9  |   |   end
10 |   end
11 end
12 else
13 |   if received message is a task assignment then
14 |   |   if robot has won several tasks then
15 |   |   |   keep the task with the best bid
16 |   |   |   Add the task to the list of tasks to announce
17 |   |   end
18 |   |   else
19 |   |   |   Keep the assigned task
20 |   |   end
21 |   end
22 end
```

Figure 7.2a and b show the difference in cost between the Hungarian method and the DMB algorithm in the case where the number of robots and tasks is equal to 10. In most cases, the DMB algorithm produces non-optimal solutions due to the fact that each robot works independently of the others.

In the DMB algorithm, each robot considers only its own benefit without taking into account the benefit of the whole system. There are some situations where a robot cannot win the best task. Consider the scenario shown in Fig. 7.2b. For example, robot $R5$ is not assigned to the best task. In order to skip this drawback, an improvement process is added to the DMB algorithm.

7.3.2 Improvement Step

The DMB consists of three main phases: announcement, bidding, and assignment. In order to make improvement, we proposed a new phase called swap-Tasks. This step consists in swapping tasks between robots in order to minimize the total cost of the whole assignment (i.e., minimize the TTD). In this step, communication between robots is required. Each robot communicates with the other robots and asks for swapping.

With regard to communication complexity, each robot needs to send $(n - 1)$ messages where n is the total number of robots. Thus, the total number of messages

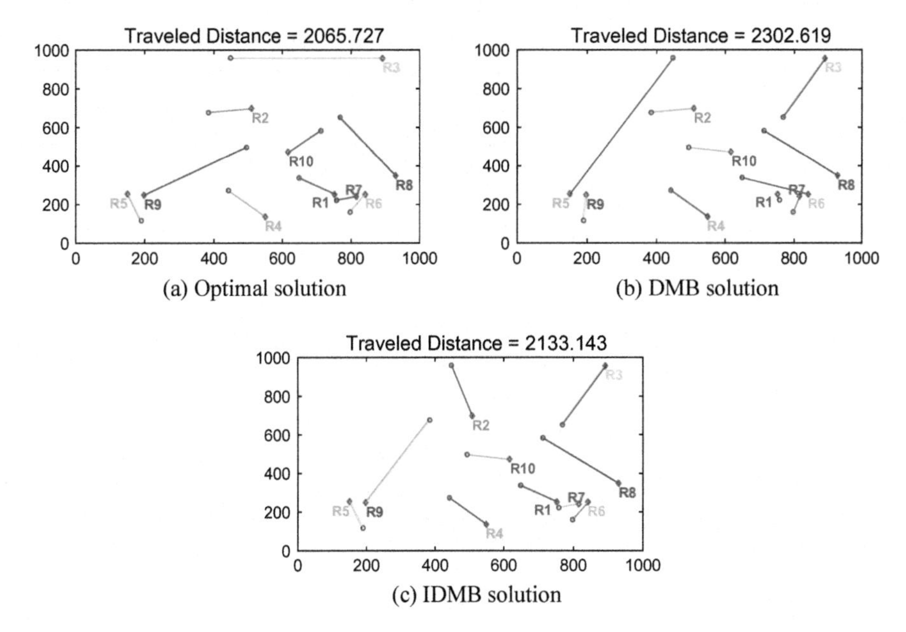

Fig. 7.2 Difference in cost between the solutions obtained with **a** the Hungarian algorithm, **b** the DMB algorithm, and **c** the IDMB algorithm. Blue squares represent the robots and red circles represent the target locations to be visited

sent by n robots is equal to $n * (n - 1)$. The swap function is repeated n times. It follows that the communication complexity is $O(n^3)$.

Assuming that the robot r_i tries to swap its task t_i with the robot r_j which is assigned to the task t_j. The robot r_i should compare the two following expressions:

$$C(t_i, r_i) + C(t_j, r_j) \tag{7.5}$$

$$C(t_i, r_j) + C(t_j, r_i) \tag{7.6}$$

where $C(t_i, r_i)$ is the cost to execute task t_i by robot r_i. Note that $C(t_i, r_i) = distance(t_i, r_i)$. If Eq. (7.5) < (7.6), then the robot r_i keeps its task t_i else robots r_i and r_j exchange their tasks. Figure 7.2b and c represent an example to assign 10 robots to 10 tasks. It is observed that the IDMB algorithm gives better results than the DMB.

7.4 Clustering Market-Based Coordination Approach

In this section, we address the problem of multi-robot systems in emergency response applications, where a team of robots/drones has to visit affected locations to provide rescue services. We formulate the problem as a multiple depot MTSP. We are interested to solve the MTSP while meeting three requirements (i) optimizes multiple objectives, including total traveled distance, maximum traveled distance, and mission time for a team of robots; (ii) provides an efficient solution; and (iii) ensures low execution times and reduces the complexity of the problem even when it scales. In fact, we considered a multi-objective problem where we need to optimize multiple objectives. We seek to find solutions that keep up the trade-off between the objectives. We propose a clustering market-based approach to solve the MTSP The approach incrementally improves the assignment and provides a solution that optimizes the conflicting objectives.

The CM-MTSP solution consists in a hybrid approach for solving the multiple depot MTSP problem that combines a clustering technique with a market-based approach with the objective of minimizing the TTD, the MTD, and the mission time. We define two main roles for the agents (i.e., the server and the robots): *auctioneer* and *bidders*. The auctioneer agent is responsible for announcing tasks and assigning each task to the agent with the best bid. In this work, a central server acts as an auctioneer and the robots as the bidders. We assume that the server initially identifies m target locations to be visited by the robots.

7.4.1 CM-MTSP Algorithm Steps

The CM-MTSP algorithm includes three steps: the clustering step, the auction-based step, and the improvement step as illustrated in Algorithm 10.

Algorithm 10: The CM-MTSP Algorithm

Input : Robots r_i $(1 < i < n)$, Targets t_j $(1 < j < m)$
Robots speed S_{r_i} $(1 < i < n)$

1 Clustering step
2 Auction-based step
3 Improvement step

Output: Assignment (r_i, c_j) $1 < i, j < n$, TTD, MTD, mission time

7.4.1.1 Clustering Step

The server first provides n clusters of locations to be visited such that the number of targets in each cluster is equal as much as possible (Algorithm 11). We used the K-means technique which is one of the most popular methods used to solve clustering problems [7]. K-means provides a partition in which elements in the same cluster are as close as possible and as far as possible from the elements in the other clusters.

Algorithm 11: Clustering step

Input : Robots r_i $(1 < i < n)$, Targets t_j $(1 < j < m)$
1 Build clusters using k-means clustering method
Output: Clusters c_i $(i < 1 < n)$

7.4.1.2 Auction-Based Step

The market process begins with an announcement phase. After forming n clusters of targets, the server announces the clusters, one by one. Note that the auction for the second cluster does not begin until the auction for the first cluster is completed. Each robot computes its bid for the announced cluster and submits this bid to the server. The cost of a robot to bid for a cluster is defined as the time necessary for that robot to visit all locations in that cluster following Eq. 7.4 and return to its initial location. In the case where a robot has a previously assigned cluster, it can ask to exchange its cluster if it discovers that the new cluster being auctioned has a lower cost to what was assigned to it. In the winner-determination stage, the server evaluates the received bids and allocates the cluster to the robot which leads to get the lowest total cost. Suppose that the server receives an exchange message from a robot which provides the best bid. In this case, the server assigns the announced cluster to that robot and adds the old cluster to its non-allocated list of clusters. This process is repeated until all clusters are allocated to the robots (Algorithm 12).

7.4.1.3 Improvement Step

The improvement step consists in the permutation of clusters between robots in order to provide a good assignment solution that simultaneously optimizes the TTD, the MTD, and the mission time (Algorithm 13). This step includes two substeps. The first consists in minimizing the mission time that results from the auction-based step. The server selects the robot that provides the maximum time and searches for the permutation that leads to decrease that time. This step is repeated while there is an improvement in terms of mission time. The second substep leads to optimize the MTD while conserving or even minimizing the mission time. The server selects the robot that provides the maximum traveled distance and searches for the permutation that leads to decrease that distance with respect to the mission time. This step is repeated while there is improvement in the MTD.

7.4.2 Illustrative Example

Consider a system with two robots (r_1 and r_2) and five targets (t_1, t_2, t_3, t_4, and t_5) as shown in Fig. 7.3. We define the speed of robot r_i as S_{r_i}. For each robot, the speed

Algorithm 12: Auction-based step

```
Input  : Robots r_i (1 < i < n), Targets t_j (1 < j < m)
         Robots speed V_{r_i} (1 < i < n)
         Clusters c_i (1 < i < n)
1  for each cluster c_i do
2      for each robot r_i do
3          Robot r_i computes cost(r_i, c_i)
4          if list_targets(r_i) = Ø then
5              robot r_i sends a bidding_msg(c_i)
6          end
7          else
8              if (cost(r_i, c_i) < cost(r_i, old_cluster(c_j))) then
9                  robot r_i sends a exchange_msg(c_j)
10                 robot r_i sends a bidding_msg(c_i)
11             end
12             else
13                 robot r_i ← c_j
14             end
15         end
16     end
17     while received_msg = true do
18         if exchange_msg(c_j) then
19             if TTD_{r_i}(c_i) < TTD_{r_i}(c_j) then
20                 r_i ← c_i, list_unassigned_clusters ← c_j
21             end
22             else
23                 best_robot ← c_i
24             end
25         end
26     end
27 end
Output: Assignment (r_i,c_j) (1 < i, j < n), TTD, MTD, mission time
```

is generated randomly in the range [0, 10]. Note that in this example $S_{r_1}>S_{r_2}$. The set of targets is decomposed into two clusters: c_1 (t_2 and t_5) and c_2 (t_1, t_3 and t_4). The bidding cost for each cluster is calculated using Eq. 7.4. Table 7.1 shows the cost calculation obtained for each robot. The server starts an auction for the cluster c_1, and both r_1 and r_2 send their bids. The server assigns c_1 to r_1 ($T_{r_1}(c_1) < T_{r_2}(c_1)$) and then makes an auction for c_2. As $T_{r_1}(c_1) < T_{r_1}(c_2)$, the robot r_1 keeps its cluster c_1 and c_2 will be assigned to the robot r_2 (Fig. 7.3a). In the improvement step, the server tries to improve the assignment. It is clear that the permutation of clusters between robots r_1 and r_2 minimizes the time cost, so c_1 will be assigned to r_2 and c_2 will be assigned to r_1 (Fig. 7.3b).

Algorithm 13: Improvement step

```
1  Function improve-time-cost  Input : Robots r_i (1 < i < n), Targets t_j (1 < j < m),
         Assignment (r_i,c_j) (1 < i, j < n)
2  while improve_time = true do
3  |  Select c_max of r_max with T_max
4  |  for r_i ≠ r_max do
5  |  |  if (cost(r_i, c_max) < T_max)&(cost(r_max, c_k) < T_max) then
6  |  |  |  r_i ← c_max, r_max ← c_k
7  |  |  end
8  |  end
9  end
   Output: Assignment (r_i,c_k) (1 < i, k < n), TTD, MTD, mission time
10 Function improve-distance-cost  Input : Robots r_i (1 < i < n), Targets t_j (1 < j < m),
         Assignment (r_i,c_k) (1 < i, k < n)
11 while improve_distance = true do
12 |  Select c_max of r_max with MTD
13 |  for r_i ≠ r_max do
14 |  |  if (cost(r_i, c_max) < T_max)&(cost(r_max, c_k) < T_max)
15 |  |        &(distance(r_i, c_max) < MTD)&(distance(r_max, c_k) < MTD) then
16 |  |  |  r_i ← c_max, r_max ← c_k
17 |  |  end
18 |  end
19 end
   Output: Assignment (r_i,c_k) (1 < i, k < n), TTD, MTD, mission time
```

Table 7.1 Bids on clusters c_1 and c_2 in terms of time

	$T(c_1)$	$T(c_2)$
Robot r_1	24.7352	32.1570
Robot r_2	125.9122	156.8610

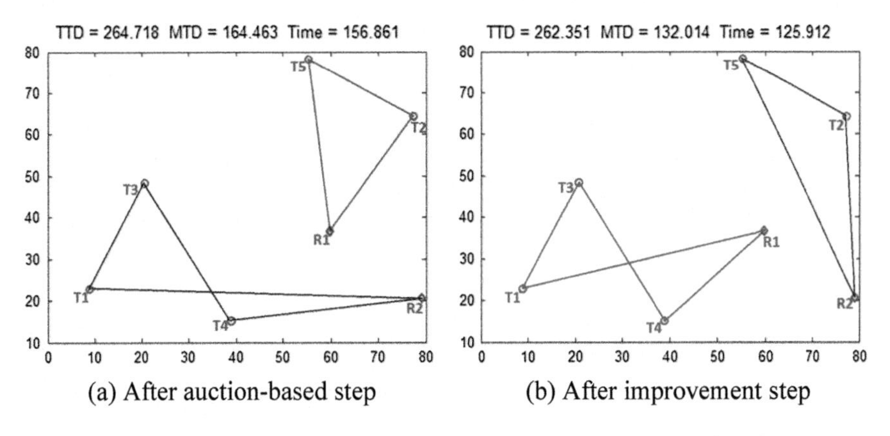

(a) After auction-based step (b) After improvement step

Fig. 7.3 Illustrative example. 2 robots (blue squares) and 5 target locations (red circles)

7.5 Fuzzy Logic-Based Approach

Fuzzy logic was introduced by Lotfi Zadeh [8] in 1965. Fuzzy logic is a mathematical discipline used to express human reasoning in a rigorous mathematical notation. A fuzzy logic system consists of four modules: fuzzification, defuzzification, inference, and rule base [9]. The fuzzification consists in transforming crisp value into fuzzy value. The crisp input is transformed into linguistic value. The membership function is used to quantify a linguistic value. The value of a membership function varies between 0 and 1.

In this section, we designed a new objective function that combines two metrics: the total traveled distance and the maximum traveled distance. Thus, the mission cost is considered as a combination of the total traveled distance and the maximum traveled distance metrics. We resort to fuzzy logic to express individual metrics in linguistic terms and then use fuzzy algebra to combine them into a crisp value that represents the degree of membership of the particular solution into the fuzzy subset of good solutions (with Short-TTD and Short-MT). A new centralized, yet fast, approach, called FL-MTSP, that takes as input the set of robots and their initial depots, and a set of target locations, and produces optimized tour assignment for each robot over a certain number of allocated target locations is designed.

7.5.1 *Fuzzy Logic Rules Design*

We propose the use of fuzzy logic [8, 9] to combine conflicting objectives. We consider the following objectives: MT and TTD, which are the inputs of the system. The proposed solution attempts to simultaneously minimize the total traveled distance and the maximum traveled distance. We assume that the mission time is proportional to the MT as the velocities of all robots are the same. In addition, we assume that the required time to collect sensor data is the same at all target locations. We rely on the expressive power of fuzzy logic to state the desired objectives to optimize. Recall that we seek to distribute a number of targets on a number of robots while minimizing the sum of all robots tour lengths as well as the maximum among all robot tour lengths. Hence, we seek solutions with Short-TTD and Short-MT. In fuzzy logic, Short-TTD and Short-MT represent fuzzy linguistic values for the fuzzy variables TTD and MT.

7.5.1.1 Linguistic Variables

In fuzzy logic, a variable could belong to a set of non-numeric values. Linguistic variables are input and output variables in the form of simple words or sentences rather than numerical values. For example, the linguistic variable of the TTD is a number of two sets according to its value: "short" and "long". As shown in Fig. 7.4a,

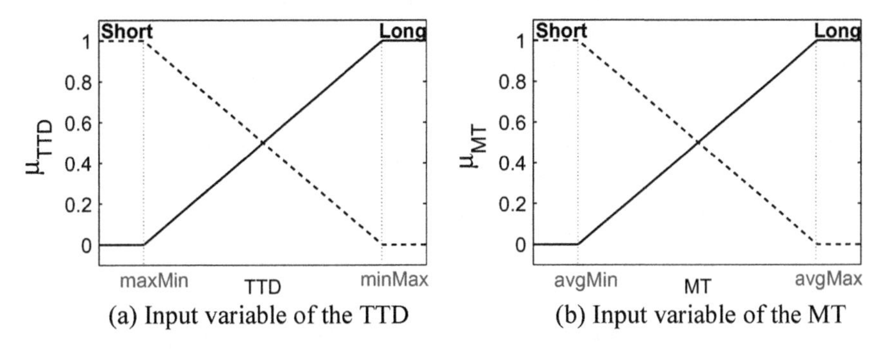

Fig. 7.4 Definition of membership functions of the inputs fuzzy sets

for values of TTD below $maxMin$, the membership to the fuzzy subset total distance that is "short" is 1. Starting from $minMax$ value, the total distance is considered totally out of the fuzzy subset of the short distances. For values of total distance between $maxMin$ and $minMax$, the membership decreases linearly from 1 to 0. The same reasoning holds for "long" distances. In a same manner, we classified the MT into two linguistic variables as shown in Fig. 7.4b.

7.5.1.2 Membership Functions

The membership function is a graphical representation used to quantify a linguistic variable. To each linguistic value corresponds a fuzzy subset with an associated membership function denoted typically by the Greek letter μ. For example, in our case μ_{TTD} gives for each solution with a given TTD (the base value) the degree of membership of that solution in the fuzzy subset of solutions with short TTD. The desire to seek solutions with simultaneous short TTD and short MT can be described by the following fuzzy rule:

IF Solution is Short-TTD **AND** solution is Short-MT **THEN** Good-Solution.

In fuzzy logic, Good-Solution is a linguistic value for the fuzzy variable solution. According to the above fuzzy rule, the membership function (μ_{GS}) of Good-Solution is expressed as follows:

$$\mu_{GS}(s) = min(\mu_{Short-TTD}(s), \mu_{Short-MT}(s))$$

The above expression assumed the minmax logic of Zadeh, where the fuzzy **AND** is interpreted as min and the fuzzy **OR** as a max.

In Fig. 7.4, the membership functions of TTD and MT variables are plotted. The membership function of each objective is determined by two thresholds. We define the shortest tour of a robot as the tour length obtained by visiting all target location using a greedy algorithm that selects the closest next target to the current position of the robot. We define the longest tour of a robot as the tour length obtained by visiting

Table 7.2 Fuzzy rules base

TTD	MT	Solution
Short	Short	Good
Long	Long	Bad
Short	Long	Bad
Long	Short	Bad

all target locations using a greedy algorithm that selects the farthest next target to the current position of the robot. We define four variables:

- $maxMin$ is defined as the longest tour among all the shortest tours of all robots going through all target locations considering our definition above.
- $minMax$ is defined as the shortest tour among all the longest tours of all robots going through all target locations considering our definition above.
- $avgMin$ as the average of the shortest tour length of all robots going through all target locations.
- $avgMax$ as the average of the longest tour length of all robots going through all target locations.

The $maxMin$ and $minMax$ are used as thresholds of the TTD fuzzy variable, because they represent reasonable upper and lower bounds on the total tour. On the other hand, $avgMin$ and $avgMax$ are used as thresholds of the MT fuzzy variable because they represent reasonable bounds of the maximum tour length (Fig. 7.4b). For example, for μ_{TTD}, for a value of TTD below $maxMin$, the total tour cost is considered short compared to the value of TTD above $minMax$. The same reasoning holds for μ_{MT}. The threshold values can be obtained using a simple greedy algorithm or using an existing TSP solver based on genetic algorithm [10]. In our solution, we have used a greedy algorithm to compute $maxMin$, $minMax$, $avgMin$, and $avgMax$.

7.5.1.3 Fuzzy Rules

In the inference phase, we define the fuzzy rules that allow the combination of our objectives which represent the inputs of our system. Each of the metrics mentioned in the previous section is considered as a different fuzzy variable. The fuzzy system has two inputs with two membership functions for each input. The rule base consists of $2^2 = 4$ rules as shown in Table 7.2. The column "solution" indicates the output fuzzy variable i.e., the quality of the assignment. The assignment can be either "Good" or "Bad" according to the TTD and MT metrics. As mentioned above, a "Good" solution is the solution with "Short" TTD and "Short" MT. We use the Mamdani Fuzzy model [11] to represent the "if-then" rules (Table 7.2).

7.5.1.4 Defuzzification

After the establishment of the rules, we move to the defuzzification step which computes the output value. We have chosen the simplest and the most commonly used method in the literature, which is the centroid defuzzification method [12]. The output value is calculated by the following equation:

$$CrispOutput = \frac{\sum_{i=1}^{N} W_i * \mu_A(W_i)}{\sum_{i=1}^{N} \mu_A(W_i)} \tag{7.7}$$

where N is the number of rules generated in the inference step (in our case N = 4), W_i is the input value, and $\mu_A(W_i)$ is the membership function of rule i.

7.5.2 *Algorithm Design*

In general, the goal of solving the MTSP is to find an optimal order to pass through all locations in order to minimize the total traveled distance. For the case where several objectives must be optimized, the goal is to find trade-off solutions while optimizing multiple performance criteria. The basic idea behind the FL-MTSP algorithm is that we rely on the fuzzy logic to combine conflicting objectives (i.e., TTD and MT). Therefore, the multi-objective MTSP is reduced to a single objective optimization problem. The objective function of the problem, which is obtained by combining the objectives to be minimized (TTD and MT), is expressed according to Eq. 7.7.

The FL-MTSP algorithm (Algorithm 14) consists of two main phases: the assignment phase (Algorithm 15) and the tour construction phase (Algorithm 16). It should be pointed out that we assume the offline version of the problem where the number of locations to be monitored is known from the beginning.

Algorithm 14: The FL-MTSP Algorithm

Input : nbTargets: number of targets
nbRobots: number of robots
matrixDistances: matrix of distance between

1 Assignment phase
2 Tour construction phase

Output: tour: tour of each robot
tourCost: tour cost of each robot
MT: max tour cost among all tours cost

7.5.2.1 Assignment Phase

The basic steps of the algorithm of the assignment is given in Algorithm 15. First, we select each target location from t_1 to t_m (m is the number of targets) and we calculate the cost of the tour for each robot if this target is assigned to it, using a greedy algorithm. The reason of using a greedy algorithm instead of a TSP solver to estimate the tour length is that the TSP solver typically takes very long time when the system becomes large, so it is limited in terms of scalability, but it can still be used for small size instances. Then, we determine the TTD cost (Eq. 7.1) and the MT cost (Eq. 7.3), which will be used as input to the fuzzy logic system to determine the membership function that represents the combination of both objectives. Then, the target location is assigned to the robot that produces the minimum value of the membership function. If multiple robots have the same fuzzy output, then we assign the target to the closest robot to the target location that leads to obtain the minimum total tour cost. This process is repeated until all targets are assigned to the robots. To improve the assignment and balance the workload between all robots, an improvement process is included in the assignment step (i.e., Line 13–22). The improvement concerns the number of targets allocated to each robot. The key idea is to change the assignment and reallocate some tasks to robots. Every robot, which has assigned to a number of target locations higher than the ratio of m by n (m/n), is selected for the improvement, and the farthest target from its assigned targets will be allocated to the nearest robot different from that robot. The Euclidean distance is used for the selection of both the farthest target and the nearest robot.

7.5.2.2 Tour Construction Phase (Algorithm 16)

Once the process of allocating targets to robots is completed, an obvious question that arises is how to construct the optimal tour of each robot. In fact, this problem is equivalent to the TSP problem. We rely on the use of a TSP solver to determine the optimal tour for each robot based on the target locations assigned to it in the previous phase. For this, we applied an existing genetic algorithm TSP solver [10] with a population size equal to 100 and a number of iterations equal to 10000.

A sample illustration of the FL-MTSP solution is given in Fig. 7.5. After assigning all target locations to the robot (Fig. 7.5b), the assignment will be modified and the robot $R5$ leaves targets $T5$, $T7$, and $T8$ and the robot $R2$ takes them (Fig. 7.5c). After applying a TSP GA solver, the tour of each robot is constructed (Fig. 7.5d).

Algorithm 15: The assignment phase

```
   Input  : nbTargets, nbRobots, matrixDistances
 1 for each target t_i do
 2   for each robot r_i do
 3     Calculate the tourCost for the robot r_i when t_i is assigned to r_i using a greedy
       algorithm
 4     Compute the TTD when t_i is assigned to r_i
 5     Apply the fuzzy logic system for TTD and MT
 6   end
 7   Select the best output obtained by the fuzzy logic system
 8   if multiple robots has the same best output then
 9     Select the nearest robot that leads to obtaining the minimum total tour cost
10   end
11   Add target t_i to the tour list of the best robot
12 end
13 for each robot r_i do
14   while the length of the allocated targets list of r_i > (m/n) do
15     Select the farthest target t_i of robot r_i from its allocated targets list
16     Find the nearest robot r_j from t_i
17     if r_j ≠ r_i then
18       Add t_i to the allocated targets list of r_j
19       Remove t_i from the allocated targets list of r_i
20     end
21   end
22 end
   Output: tour of each robot
```

Algorithm 16: Tour construction phase

```
  Input  : nbRobots
           tour: tour of each robot after the assignment step
1 for each robot r_i do
2   Apply the TSP_GA solver
3 end
  Output: tour, tourCost
```

7.6 Move-and-Improve: A Market-Based Multi-robot Approach for Solving the MD-MTSP

The Move-and-Improve mechanism comprises four phases: (1) the allocation phase, (2) the tour construction phase, (3) the overlapped targets elimination phase, and (4) the solution improvement phase (see Fig. 7.6 and Algorithm 17). A video demonstration of the Move-and-Improve mechanism using Webots simulator is available in [13].

In the allocation phase, presented in Algorithm 18, each robot receives the list of available targets to be visited. This list of targets can be sent, for example, by a control station. Each robot R_i maintains two lists: (*i*) The list of *available* targets that are

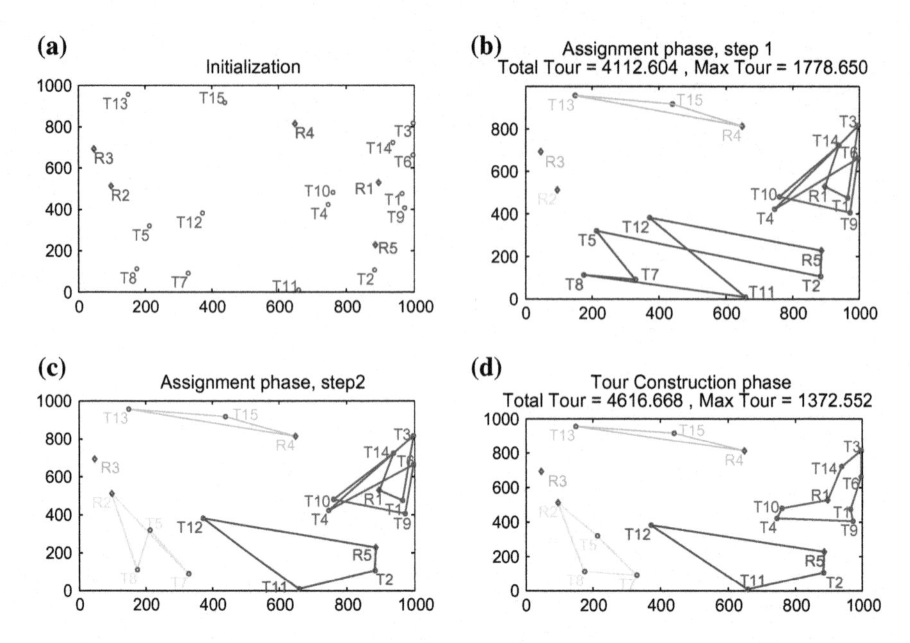

Fig. 7.5 Simulation example with 5 robots and 15 target locations. **a** Initial position of the robots and the targets to be allocated. The blue squares represent the robots and the red circles represent the target locations. **b** Tour of each robot after applying the fuzzy logic approach. **c** Final assignment after redistributing the targets. **d** Final tour of each robot after applying the TSP GA solver [10]

Algorithm 17: Move-and-Improve General Algorithm

Begin
Target Allocation
Tour Construction
Overlapped Targets Elimination
Solution Improvement
End

not already allocated, and (*ii*) the list of *allocated* targets, which contains the targets visited and allocated to the robot R_i itself. At the start of the mission, all targets are marked as *available* in each robot, and the list of allocated targets is empty. Then, each allocated target will be removed from the available targets list, and the robot informs its neighbors to also remove it from their available targets list. We note that the target becomes allocated to the robot R_i when it reaches the target. We also refer to an allocated target as a *visited* target. Each robot starts by computing the cost (e.g., the traveled distance) of visiting each available target and finally selects the target with the lowest cost (e.g., the nearest target). Then, the robot R_i starts moving toward the selected target $T_{Selected}$, and in the meanwhile, it keeps discovering other robots in its neighborhood within its communication range and exchanging information about allocated targets. Indeed, when a robot discovers another robot in its vicinity, they

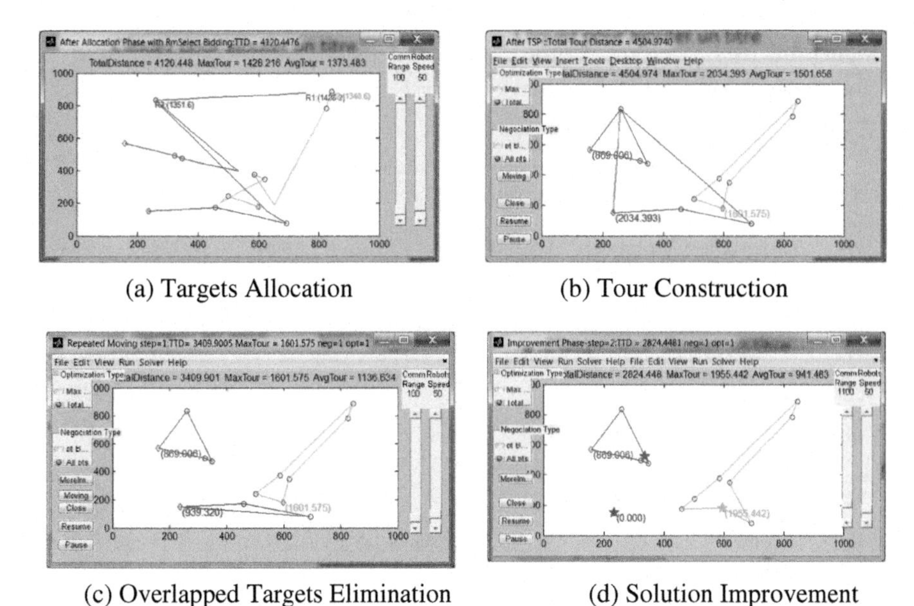

(a) Targets Allocation

(b) Tour Construction

(c) Overlapped Targets Elimination

(d) Solution Improvement

Fig. 7.6 Move-and-Improve

both exchange their lists of available targets that are not yet visited nor allocated. As such, each robot can update its own list of available targets by discarding those allocated/visited by other robots. In addition, if the currently selected target is found to be no longer available (i.e., it was already allocated to another robot), the robot dismisses that target and looks for another one (Algorithm 18: Line 10). If the selected target $T_{Selected}$ is still not allocated to other robots, the robot bids on this target with its neighbor robots. The neighbor robots will send their costs to the robot R_i, which played the auctioneer role for target $T_{Selected}$, in response to the bidding request. Finally, the robot R_i will assign the target $T_{Selected}$ to the robot with the lowest cost, including itself (Algorithm 18: Line 14).

This process of moving toward the selected target $T_{Selected}$ and exchanging available targets is continuously repeated until the robot R_i dismisses the selected target when it discovers that it is allocated to another robot, or until it reaches the selected target. In the latter case, it adds this target to its own list of allocated targets and removes it from its own list of available targets. The robot R_i repeats the process of selecting a new target and moving to it until the list of available targets becomes empty, as illustrated in Algorithm 18.

Upon the completion of the allocation process (the list of available targets becomes empty), each robot constructs an optimal tour across its allocated targets using a TSP solver (Phase 2), as illustrated in Algorithm 19. The robots can use any TSP tour construction algorithm proposed in the literature such as LKH [14] or [15]. In our MATLAB simulation, we used an existing implementation of a TSP solver based on genetic algorithms [10] for tour construction.

Algorithm 18: Phase 1-Target Allocation

```
1:  Available Targets List = All targets to be visited
2:  R_i Allocated Targets List = Ø
3: while Available Targets List ≠ Ø do
4:   Compute cost from current position to each available target
5:   Select the target with the lowest cost
6:   repeat
7:     Discover neighbor robots
8:     if neighbor ≠ Ø then
9:       Communicate with neighbor(s) and update list of available targets
10:        if selected target ∉ available targets then
11:          go to line 3:
12:          {the selected target was already allocated to an other robot}
13:        end if
14:        Bid on the selected target
15:        if cost of robot ri > cost of robot rj, where rj is a neighbor of ri then
16:          go to line 3:
17:          {the target can be visited by other robot with minimum cost}
18:        end if
19:     end if
20:     Move robot to the selected target
21:  until Robot reach the selected target
22:  Add target to the set of allocated targets
23:  Remove target from the set of available targets
24: end while
```

Algorithm 19: Phase 2-Tour Construction

```
call tsp (Allocated targets)(where tsp is a travelling salesmen algorithm)
```

After the tour construction process, each robot generates a first solution to the MD-MTSP problem and obtains a proper tour to follow through its allocated targets. However, it may happen that some targets would have been allocated to or visited by several robots during the allocation phase. This is possible as robots have limited communication ranges and may not have the opportunity to exchange their lists of allocated targets, either directly or indirectly through other neighbors.

The objective of Phase 3, presented in Algorithm 20, consists in improving the MD-MTSP solution by eliminating common targets allocated to more than one robot, through a distributed market-based approach, while the robots are moving through their constructed tours. As such, during their tour missions, when two robots are able to communicate, they exchange their lists of allocated targets, and in case there are one or more overlapped targets, the robots will bid on these targets. More precisely, considering a robot R_i and a robot R_j both have the same allocated target T_k. If the gain resulting from eliminating T_k from the tour of R_i is greater than the gain resulting from eliminating T_k from the tour of R_j, then T_k will be eliminated from the tour of R_i, and R_i computes a new tour based on its remaining allocated targets. The gain is defined as the difference between the old tour cost before removing the common target and the new tour cost after target removal. The gain differs by either we wish to minimize the total cost (the sum of traveled distance) or to minimize the maximal individual cost (the maximum tour). Moreover, in case there are several overlapped targets, the robots can bid on them separately (one by one) or all together. Returning to the previous example, where we wish to minimize the total cost, we define:

$$\begin{aligned} gain_i &= C(Tour_{R_i}) - C(Tour_{R_i} \backslash T_k) \\ gain_j &= C(Tour_{R_j}) - C(Tour_{R_j} \backslash T_k) \end{aligned}$$

where $C(Tour_{R_i})$ is the cost of $Tour_{R_i}$ containing the target T_k and $C(Tour_{R_i} \backslash T_k)$ is the cost of $Tour_{R_i}$ after removing the target T_k. $C(Tour_{R_i} \backslash T_k)$ is computed using the TSP algorithm applied on the targets allocated to R_i, except the target T_k. $gain_i$ represents the gain obtained when we remove target T_k from the tour of robot R_i and $gain_j$ represents the gain obtained when we remove target T_k from the tour of robot R_j. If $gain_i > gain_j$, it is more beneficial to remove T_k from the tour of robot R_i and keep it in the tour of robot R_j. If the objective function was to minimize the maximum cost, the gain will be defined as follows:

$$\begin{aligned} gain_i =& \max(C(Tour_{R_i}), C(Tour_{R_j})) \\ &- \max(C(Tour_{R_i} \backslash T_k), C(Tour_{R_j})) \\ gain_j =& \max(C(Tour_{R_i}), C(Tour_{R_j})) \\ &- \max(C(Tour_{R_i}), C(Tour_{R_j} \backslash T_k)) \end{aligned}$$

The last phase (the solution improvement phase), illustrated in Algorithm 21, consists in looking for possible additional optimization of the tours resulting from Phase 3. For that purpose, when a robot R_i enters in the communication range of a robot R_j, it sends to R_j a bidding request on the target T_k that induces the biggest cost to visit in its tour. In other words, T_k is the target that when we remove from the tour of robot R_i we get the minimal cost, compared to other allocated targets. Formally speaking, T_k can be computed as:

$$T_k = \arg\min_{T_j} C(Tour_{Ri} \backslash T_j)$$

The robot R_i computes the gain obtained when it removes this target T_k from its tour, and the neighbor robot R_j calculates the extra cost obtained when it adds this target T_k to its tour. In case of *MinSum*, the gain is computed as follows:

$$\begin{aligned} gain =& [C(Tour_{R_i}) + C(Tour_{R_j})] \\ &-[C(Tour_{R_i} \backslash T_k) + C(Tour_{R_j} \cup T_k)] \end{aligned}$$

$C(Tour_{R_j} \cup T_k)$ is computed using the TSP algorithm applied on the targets allocated to R_j union T_k. If the gain obtained by removing T_k from the R_i's tour is greater than

Algorithm 20: Phase 3-Overlapped Targets Elimination

```
TourDone=false {indicates if a total tour is done}
current position=depot {position the robot to the first target of the tour, i.e, the depot }
repeat
  Search for neighbor robots
  if neighbors ≠ ØØ then
    Exchange list of Allocated targets with neighbors
    Common targets=Allocated targets ⋂(∪{neighbors' Allocated targets})
    for all ti ∈ common targets do
      Bid on ti
      if I am not the winner of ti then
        remove ti from Allocated target
        call tsp (Allocated targets) {construct the new tour}
        current position=depot
        TourDone=false
      end if
    end for
  end if
  Move robot to next target
  if current position==depot then
    TourDone==true
  end if
until TourDone
```

the extra cost obtained when adding T_k to R_j's tour, the neighbor robot R_j wins this target T_k, and therefore, T_k will be removed from robot R_i's allocated targets and added to the list of R_j's allocated targets. In case of *MinMax*, the gain is computed as follows:

$$
\begin{aligned}
gain = & \max(C(Tour_{R_i}), C(Tour_{R_j})) \\
& - \max(C(Tour_{R_i} \backslash T_k), C(Tour_{R_j} \cup T_k))
\end{aligned}
$$

7.7 Conclusion

In this chapter, we have presented four different approaches for the MRTA problem: (1) improved market-based approach, (2) clustering market-based approach, (3) fuzzy logic-based approach, and (4) Move-and-Improve approach. The improved market-based approach allows robots to communicate and coordinate together in order to increase the possibility of obtaining an optimized assignment of tasks to robots. The reallocation is used to ensure the condition of only assigning one task to a robot. After assigning each task to a robot, an improvement step is applied. It consists in swapping tasks between robots in order to reduce the cost of the assignment. The clustering market-based approach is based on the use of a clustering method with an auction process. We considered a multi-objective problem where we seek

Algorithm 21: Phase 3- Solution Improvement

```
TourDone=false {indicates if a total tour is done}
current position=depot {position the robot to the first target of the tour, i.e, the depot }
repeat
  Search for neighbor robots
  if neighbors ≠ Ø then
    select the worst allocated targets (tw) (the one that have the worst cost).
    Bid on tw
    if I am not the winner of tw then
      remove tw from Allocated target
      call tsp (Allocated targets) {construct the new tour}
      current position=depot
      TourDone=false
    end if
  end if
  Move robot to next target
  if current position==depot then
    TourDone==true
  end if
until TourDone
```

to find solutions that keep up the trade-off between the objectives. In addition, we aim to uniformly distribute targets between robots such that the number of allocated targets for each robot is equal or close. The fuzzy logic-based approach is a centralized solution based on the use of the fuzzy logic algebra to combine two objectives: the objective of minimizing the total traveled distance by all the salesmen and the objective of minimizing the maximum traveled distance by any robot. The approach consists of two phases: the assignment phase where the targets allocation is based on the output of the fuzzy logic system, and the tour construction phase, where we used an existing genetic algorithm to build a suboptimal tour for each robot. The Move-and-Improve approach is a distributed market-based algorithm where robots progressively select and move toward the target with the lowest cost. It consists of four steps: the initial target allocation step, the tour construction step, the negotiation of conflicting targets step, and the solution improvement step.

In the next chapter (Chap. 8), a detailed performance evaluation of the approaches proposed for solving the MRTA problem will be presented.

References

1. Trigui, Sahar, Anis Koubaa, Omar Cheikhrouhou, Habib Youssef, Hachemi Bennaceur, Mohamed-Foued Sriti, and Yasir Javed. 2014. A distributed market-based algorithm for the multi-robot assignment problem. *Procedia Computer Science*, 32(Supplement C): 1108–1114. The 5th International Conference on Ambient Systems, Networks and Technologies (ANT-2014), the 4th International Conference on Sustainable Energy Information Technology (SEIT-2014).

2. Trigui, Sahar, Anis Koubâa, Omar Cheikhrouhou, Basit Qureshi, and Habib Youssef. 2016. A clustering market-based approach for multi-robot emergency response applications. In *2016 international conference on autonomous robot systems and competitions (ICARSC)*, 137–143. IEEE.
3. Trigui, Sahar, Omar Cheikhrouhou, Anis Koubaa, Uthman Baroudi, and Habib Youssef. 2016. Fl-mtsp: A fuzzy logic approach to solve the multi-objective multiple traveling salesman problem for multi-robot systems. *Soft Computing*: 1–12.
4. W Kuhn, Harold. 1955. The hungarian method for the assignment problem. *Naval Research Logistics (NRL)* 2 (1–2): 83–97.
5. Bernardine Dias, M., Robert Zlot, Nidhi Kalra, and Anthony Stentz. 2006. Market-based multirobot coordination: A survey and analysis. *Proceedings of the IEEE* 94 (7): 1257–1270.
6. Golmie, Nada, Yves Saintillan, and David H Su. 1999. A review of contention resolution algorithms for IEEE 802.14 networks. *IEEE Communications Surveys* 2 (1):2–12.
7. Chan, Zeke S.H., Lesley Collins, and N. Kasabov. 2006. An efficient greedy k-means algorithm for global gene trajectory clustering. *Expert Systems with Applications* 30 (1): 137–141.
8. Zadeh, L.A. 1965. Fuzzy sets. *Information and Control* 8 (3): 338–353.
9. Zadeh, Lotfi Asker. 1975. The concept of a linguistic variable and its application to approximate reasoning — I. *Information Sciences* 8 (3): 199–249.
10. Kirk, Joseph. 2011. Traveling-salesman-problem-genetic-algorithm. http://www.mathworks.com/matlabcentral/fileexchange/13680-traveling-salesman-problem-genetic-algorithm.
11. Mamdani, Ebrahim H., and Sedrak Assilian. 1975. An experiment in linguistic synthesis with a fuzzy logic controller. *International Journal of Man-Machine Studies* 7 (1): 1–13.
12. Takagi, Tomohiro, and Michio Sugeno. 1985. Fuzzy identification of systems and its applications to modeling and control. *IEEE Transactions on Systems, Man and Cybernetics* 1: 116–132.
13. Webots simulation scenarios. 2014. http://www.iroboapp.org/index.php?title=Videos.
14. Lin, Shen. 1973. An effective heuristic algorithm for the traveling-salesman problem. *Operations Research* 21 (2): 498–516.
15. Braun, Heinrich. 1991. On solving travelling salesman problems by genetic algorithms. In *Parallel problem solving from nature*, 129–133. Berlin: Springer.

Chapter 8
Performance Analysis of the MRTA Approaches for Autonomous Mobile Robot

Abstract The multi-robot task allocation is a fundamental problem in robotics research area. Indeed, robots are typically intended to collaborate together to achieve a given goal. This chapter studies the performance of the IDBM, CM-MTSP, FL-MTSP, and Move-and-Improve approaches. In order to highlight the performance of the proposed schemes, we compared each one to appropriate existing ones. IDMB was compared with the RTMA [1], CM-MTSP was compared with single-objective and greedy algorithms, and FL-MTSP was compared with a centralized approach based on genetic algorithm and with NSGA-II algorithm. To validate the efficiency of the Move-and-Improve distributed algorithm, we first conducted extensive simulations and evaluated its performance in terms of the total traveled distance and the ratio of overlaped targets under different settings. The simulation results show that IDMB and Move-and-Improve algorithms produce near-optimal solutions. Also, CM-MTSP and FL-MTSP provide a good trade-off between conflicting objectives.

8.1 Introduction

In complex robotics applications, a cooperative robot system represents a recommended alternative to a single-robot system considering the collaborative effect between robots, which leads to accomplish their mission more efficiently. Multi-robot coordination is a major challenging problem that arises in a vast array of applications. Several research works addressed this problem and proposed solutions, which can be classified from two perspectives: (1) algorithmic approaches, which can be either centralized or distributed (2) optimization problem type, which can be either single-objective or multi-objective. Centralized approaches typically rely on evolutionary algorithms [2, 3], which have the advantage of converging to good solutions, but at the cost of intensive computation requirements and long execution times. On the other hand, distributed approaches, including market-based techniques [4, 5], provide lower-quality solutions in general, but executes much faster. The complexity of these approaches significantly increases when the problem switches from single-objective to multi-objective optimization, where the goal is to optimize several metrics that can be conflicting in nature. This class of problems usually has more than

A. Koubaa et al., *Robot Path Planning and Cooperation*, Studies in Computational Intelligence 772, https://doi.org/10.1007/978-3-319-77042-0_8

one optimal solution that provides a balanced optimization of the different metrics, also known as Pareto-optimal solutions [6]. Evolutionary algorithms are the typical approaches to solve this type of multi-objective optimization problems to find the Pareto-optimal solutions. However, the execution time is too long when the problem scales, making it non-appropriate for situations where an efficient solution is needed in (near) real-time

In this chapter, we present the performance evaluation of the IDMB approach, the CM-MTSP approach, the FL-MTSP approach, and Move-and-Improve approach proposed in Chap. 7.

8.2 Performance Evaluation of the IDMB Approach

8.2.1 Simulation Study

To study the performance of the DBM and IDMB algorithms, we used a simple simulator implemented using MATLAB. The robots and tasks are dispersed in a free-obstacles environment of $1000 * 1000\,\text{m}$. The task cost is calculated as the Euclidean distance. We tested the algorithms on different cases (i.e., 2, 5, 8, 10, 12, 15, 20, and 30 robots and tasks). For each case, we considered 30 different scenarios. Each scenario is specified by the coordinates of a randomly chosen initial position of both robots and tasks. Each scenario is repeated 30 times (i.e., 30 runs for each scenario), and we recorded the global estimated cost. To obtain the optimal allocation, we used the well-known Hungarian method [7]. We also implemented the RTMA [1] proposed to solve the task formation problem, where every robot is able to be allocated to only one task. Similar to the IDMB, the RTMA algorithm is based on the market process. In order to demonstrate the efficiency of the IDMB algorithm, we considered the error in percentage in comparison with the optimal solution obtained with the Hungarian algorithm. The results are shown in Fig. 8.1. It can be observed that the solutions obtained with the IDMB are very close to the optimal (Fig. 8.2). For less than 10 robots and tasks, the algorithms (DMB, IDMB, and RTMA) give good results (the error is less than 7.8% for DMB, 0.5% for IDMB and 6.8% for RTMA). Increasing the number of robots and tasks, the IDMB algorithm still produce near-optimal solutions as shown in Figs. 8.1 and 8.2. The maximum error obtained with the IDMB algorithm in all cases did not exceed 2%. This means that the IDMB algorithm can generate the optimal solution in several simulations. Notice that in the case where the number of robots and tasks is 2, the IDMB algorithm always produces the optimal solution (error is 0%).

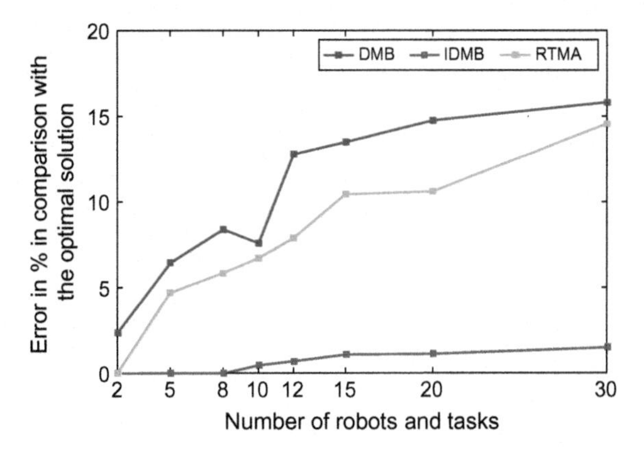

Fig. 8.1 Error in percentage in comparison with the optimal solution for the DMB, the IDMB, and the RTMA algorithms

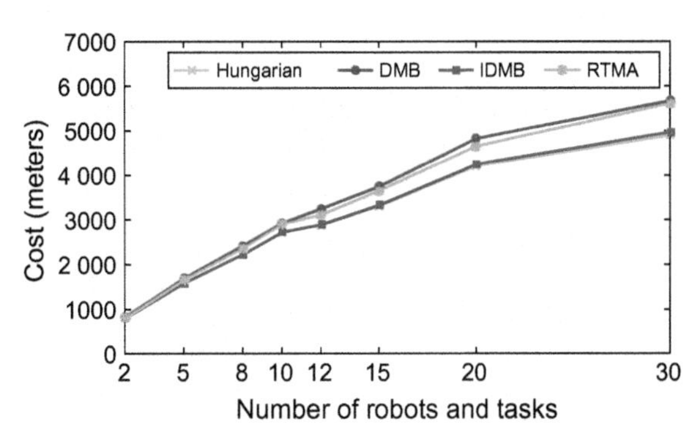

Fig. 8.2 Results of the estimated cost of the Hungarian, DMB, IDMB, and RTMA algorithms over 30 simulations per case

8.2.2 Experimentation

We implemented the proposed distributed market-based approach on real robots to demonstrate the feasibility of the algorithm in real-world deployment. We used the Turtlebot [8] V2 robots as robotic platform and the open-source robot operating system [9] to implement the robot messaging, low-level control, and task assignment. Each robot relies on the default ROS navigation stack for navigation, localization, and obstacle avoidance purposes.

The implementation mainly consists of three main components: (i) a TurtleBot controller component, which consists of a set of modular classes that provide all functionalities needed to monitor sensor data of the robot and control its motors (e.g., sending the robot a goal), (ii) an MRTA server, which is a UPD server that receives bidding and auctioning messages from neighbor robots and publishes the received message as a ROS topic, *mrta_topic*, for other ROS subscribers processing the messages, (iii) an MRTA processing node, which is a ROS node that subscribes to the *mrta_topic* topic published by the MRTA server and processes the received messages depending on their types, which are target announcement, target auctioning, target bidding, and target assignment.

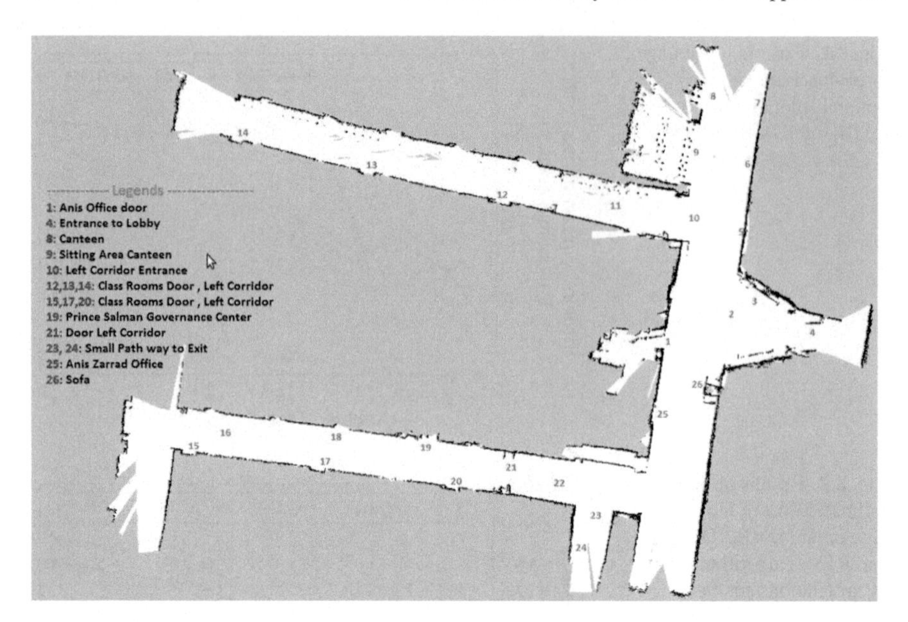

Fig. 8.3 ROS map used for experiments in Prince Sultan University

The experiments were performed in the corridors of the College of Computer and Information Sciences of Prince Sultan University (refer to map in Fig. 8.3) with two Turtlebot robots. In the experiments, we tested a simple scenario with two robots $R1$ and $R2$ that select two targets locations $T1$ (location 26 in Fig. 8.3) and $T2$ (location 4 in Fig. 8.3), respectively, such that the optimal assignment is opposite to the initial assignment. After the execution of the DBM distributed algorithm, both robots reached the consensus about the optimal assignment such that $R2$ assigns its initial $T2$ to $R1$ and $R1$ assigns its initial target $T1$ to $R2$. A video demonstration of the experiments is available on this iroboapp project Web site [10]. We used the Euclidean distance for cost estimation. More accurate costs can be obtained using the total length of the path planned by the ROS global planner.

8.3 Performance Evaluation of the CM-MTSP Approach

In this section, we present the performance evaluation of the clustering market-based approach for solving the multiple depot MTSP. We build our simulation using MATLAB. We adopted the test problems where the number of target locations varies in [20, 50, 100] and the number of robots varies in [3, 6, 12]. Robots and targets positions are placed in the range of [0, 100]. The LKH-TSP solver [11] is used to find the least distance for the robot to travel from a fixed starting point while visiting the target locations exactly once. The LKH-TSP solver has shown its ability to produce optimal solutions to most problems. Also, the LKH-TSP is tractable for large-scale problems and can generate solutions within a small execution time [12, 13]. For each scenario, we performed 30 different runs of the algorithm and each runs with different clusters. We evaluated the TTD, the MTD, and the mission time objectives. We explored the performance of the proposed algorithm by varying the number of robots and targets.

8.3.1 *Comparison of the CM-MTSP with a Single-Objective Algorithm*

The efficiency of our solution is validated through comparison with a clustering single-objective market-based algorithm (CSM-MTSP). In CSM-MTSP, the process is the same as in the CM-MTSP, but the server uses the TTD as a unique cost metric to assign clusters to robots. Figures 8.4, 8.5, and 8.6 show the total traveled distance, the maximum traveled distance and the mission time, respectively, for both the CM-MTSP and CSM-MTSP algorithms. As shown in Figs. 8.4 and 8.5, the gap between the CM-MTSP and CSM-MTSP in terms of TTD and MTD is in the range of $[\mathbf{1}, \mathbf{10\%}]$ for all scenarios in favor of the single objective. On the other hand, the gap between CM-MTSP and CSM-MTSP in terms of mission time is in the range of $[\mathbf{8}, \mathbf{35\%}]$, with a more significant gain for the multi-objective CM-MTSP. This demonstrates that CM-MTSP provides a better trade-off in satisfying the different objectives as compared to CSM-MTSP. The reason for having reduced mission time is that CM-MTSP forms clusters with the objective to group target locations close to each other so as to reduce the maximum tour of the robot and also assign clusters based on the mission time, so that slower robots are assigned to clusters with the shortest tours. In addition, the improvement phase of the CM-MTSP further optimizes the mission time. This feature is more interesting for applications with real-time constraints such as emergency response applications.

In addition, the decrease of the maximum traveled distance with the increase of the number of robots indicates that there is a uniformity of assignment that means that the number of allocated targets for each robot is equal or close. Figure 8.7 shows an example of distribution of targets between robots.

8.3.2 *Comparison of the CM-MTSP with a Greedy Algorithm*

To prove the efficiency of the clustering approach, we compare its performance against a greedy market-based algorithm that allocates targets to robots one by one without prior clustering. The concept of the greedy market-based algorithm is similar to the CM-MTSP algorithm, but we consider individual target allocations instead of cluster allocations.

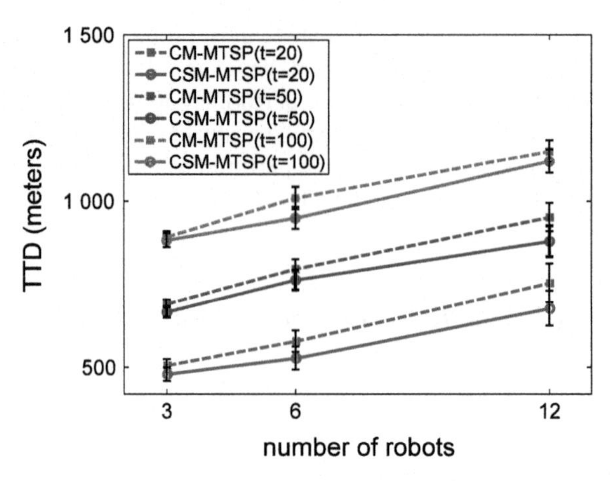

Fig. 8.4 TTD of CM_MTSP and CSM_MTSP solutions

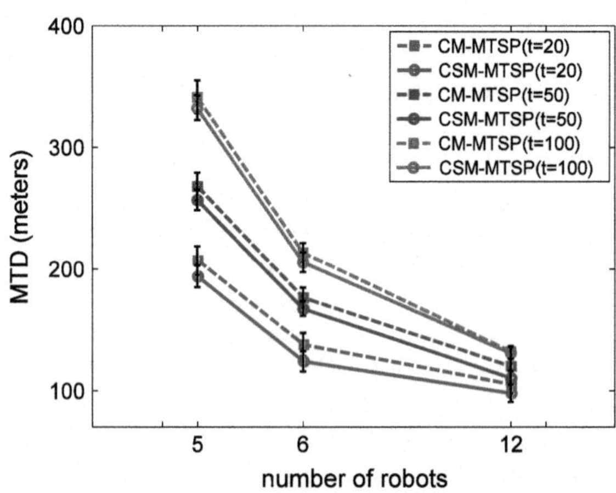

Fig. 8.5 MTD of CM_MTSP and CSM_MTSP solutions

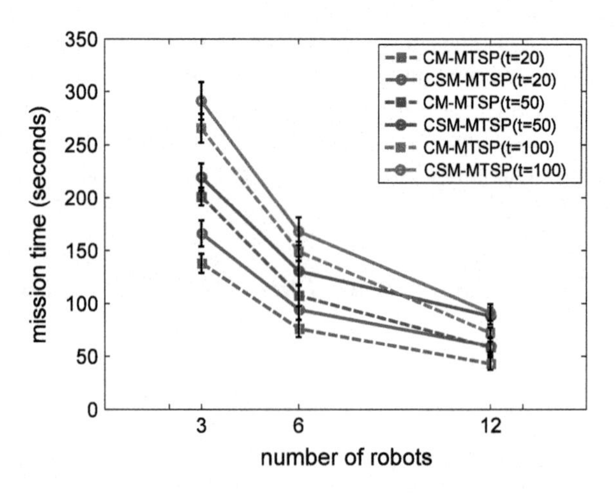

Fig. 8.6 Mission time of CM_MTSP and CSM_MTSP

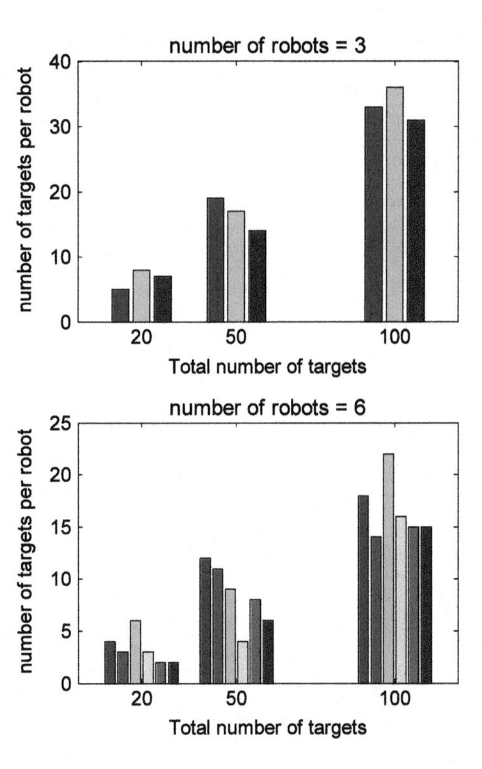

Fig. 8.7 Distribution of targets in the case of 3 and 6 robots

The algorithm works as follows. The server starts an auction for each target t_i. In the bidding phase, each robot can either bid for a new target t_i, or try to exchange its worst allocated target that leads to increase the mission time of its tour with the announced task t_i. At the end, the server assigns the target t_i to the robot with the least cost. This process is repeated until all targets are assigned to robots.

We considered different scenarios, where the number of targets varied in [20, 100], and the number of robots varied in [3, 6, 12]. We performed 10 runs for each scenario to ensure 95% confidence interval. Figure 8.8 shows the comparative results.

In Fig. 8.8d, we observe that the CM-MTSP significantly reduces the execution time as compared to the greedy algorithm. For example, in the case of 12 robots and 100 targets, the reduction exceeds 95%. This is due to the fact that we used a clustering technique to group targets. So, instead of assigning m targets, we only search to assign $n << m$ clusters. This will significantly reduce the execution time for large instances, which demonstrates that CM-MTSP scale much better than traditional non-clustered market-based approaches.

In addition, we observe that the MTD of the CM-MTSP algorithm was decreased in comparison with the greedy algorithm (Fig. 8.8b). For example, in the case of 12 robots and 100 targets, the MTD is reduced by around 30%. This means that the number of targets assigned to each robot is not the same for both the algorithms. For the greedy algorithm, not all robots are assigned to targets and so, in this case the

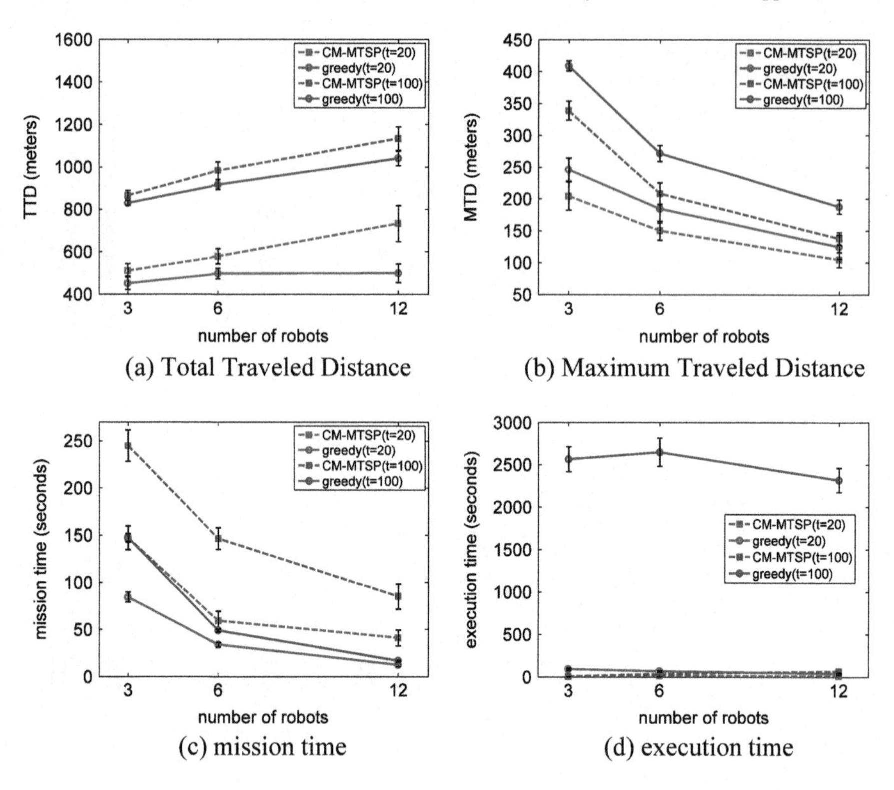

Fig. 8.8 Comparison results of the CM-MTSP with a greedy algorithm

MTD will increase. Figure 8.9 shows an illustrative example, between the greedy solution and the CM-MTSP algorithm using a scenario with 6 robots and 20 targets. The increase of the TTD of the CM-MTSP in comparison with the greedy algorithm as shown in Fig. 8.8a is attributed to the fact that the targets in the same cluster can be far from each other or even when the number of robots is high with respect to the number of targets. For example, it is clear from Fig. 8.8a that the gap between the CM-MTSP algorithm and the greedy algorithm, in the scenario with 6 robots and 20 targets, is reduced in comparison with the scenario with 12 robots and 20 targets.

The result of mission time (Fig. 8.8c) the greedy algorithm conforms the results obtained for the MTD. In the case where a small number of robots is assigned to targets, the TTD increases and the mission time decreases in contrast to the CM-MTSP algorithm.

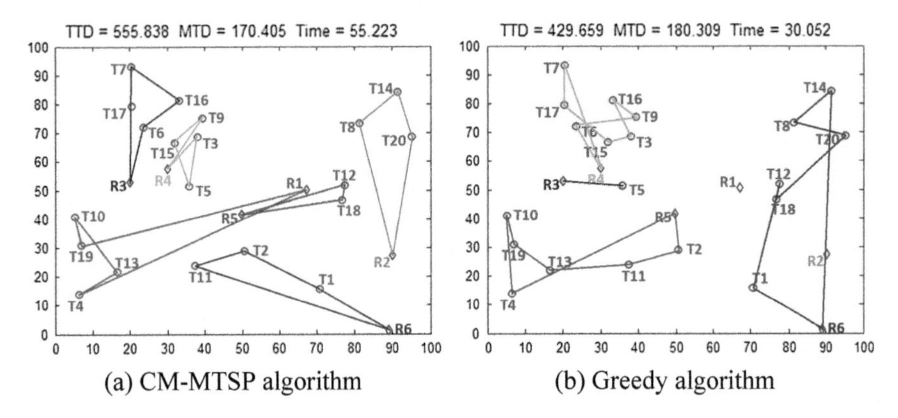

Fig. 8.9 Simulation example of the CM-MTSP and the greedy algorithm

8.4 Performance Evaluation of the FL-MTSP

We have built our own custom simulation using MATLAB under Windows OS to implement the proposed approach. All simulations are run on a PC with an Intel Core i7 CPU @ 2.40 GHz and 6 GB of RAM. We evaluate the performance of the FL-MTSP algorithm with two objective functions in scenarios without obstacles where the tour cost is calculated as the Euclidean distance. We adopted the same test problems used in [4]. The number of robots n varies in [3, 10, 20, 30, 100], whereas the number of targets locations m varies in the interval [30, 70, 100, 200, 300]. An $(m * m)$ cost matrix is randomly generated and contains the distances between targets. Targets positions are placed in the range of [0, 1000]. In addition, an $(n * m)$ cost matrix is randomly generated and contains the distances between each robot and all targets. Moreover, robots are randomly placed in the range of [0, 1000]. The GA is used to find the least distance for the robot to travel from a fixed starting point and end positions while visiting the other places exactly once. For each scenario, we performed 30 different runs for the algorithm to ensure 95% confidence interval. For each run, we recorded the tour cost for each robot, the TTD, the MT cost (i.e., the MTD) and the execution time. The execution time of the algorithm is the average of the 30 execution times. We have explored the performance of the proposed approach under varying the number of robots and targets.

8.4.1 Impact of the Number of Target Locations

Figure 8.10 shows the TTD and the MT cost as a function of the number of targets, for a fixed number of robots. The TTD is presented in Fig. 8.10a, and the MT cost is presented in Fig. 8.10b. We can observe that, in most cases, the TTD and the MT cost increase with the increase of the number of targets for a fixed number of robots.

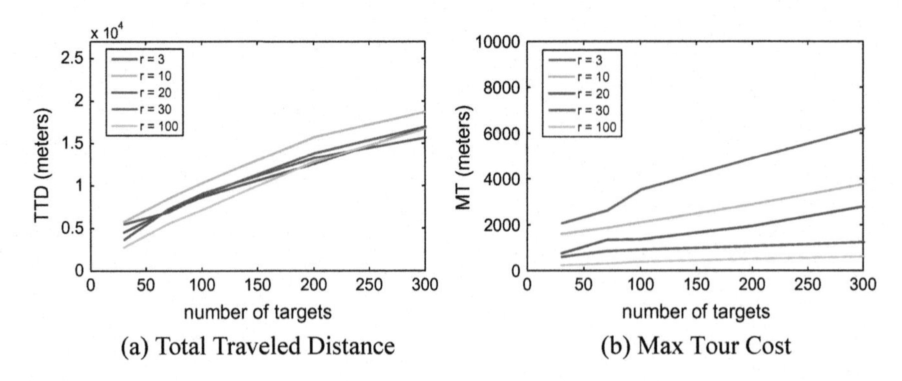

(a) Total Traveled Distance (b) Max Tour Cost

Fig. 8.10 Impact of the number of targets on the total traveled distance and max tour cost (number of robots is fixed)

The assignment becomes more difficult when more target locations are involved since the algorithm needs to determine the route for each robot while maintaining the minimum TTD and MT cost. We conclude that the increase of the number of targets affects the system performance. It is interesting to notice that increasing the number of robots does not change much the total traveled distance, while there is a huge drop in the maximum tour cost.

8.4.2 Impact of the Number of Robots

To study the impact of the number of robots, we performed simulations where we fixed the number of targets while varying the number of robots. The results are shown in Fig. 8.11. We observe that, in most cases, the total traveled distance (Fig. 8.11a) slightly decreases when the number of robots increases. Moreover, the max tour cost exponentially decreases while increasing the number of robots especially when the number of robots is less than 30 (Fig. 8.11b). This means that the target locations are shared between multiple robots in a way to decrease both the total traveled distance and the maximum tour cost. This result shows the benefit of the use of multiple robots to solve the TSP problem. However, as indicated earlier, the overall traveled distance cost is not affected much by increasing the number of robots for a given number of tasks.

8.4.3 Comparison with MDMTSP_GA

In order to highlight the performance of our solution, we compared with the MDMTSP_GA solution [14] under the MATLAB simulator. MDMTSP_GA is a

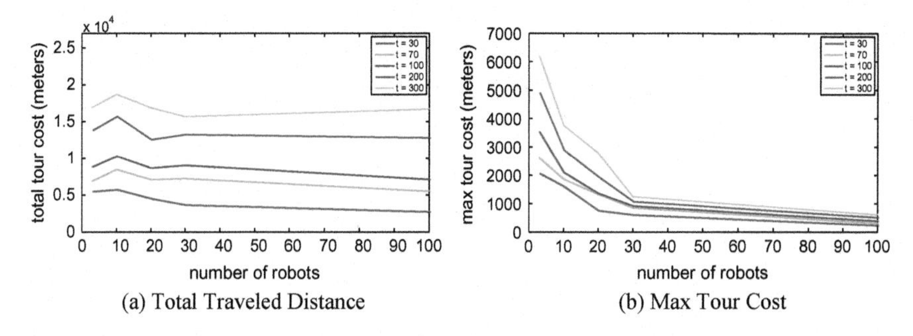

(a) Total Traveled Distance (b) Max Tour Cost

Fig. 8.11 Impact of the number of robots on the total traveled distance and max tour cost (number of targets is fixed)

centralized approach based on genetic algorithm. We used a population size equal to 240 and a number of iterations equal to 10000. These parameters are sufficient to generate good solutions. From Fig. 8.12, we observe that our algorithm outperforms the MDMTSP_GA in terms of TTD. We can notice that for a large number of robots and targets, the gap between our solution and the MDMTSP_GA solution increases. For example, when the number of targets is 200 and the number of robots is 30, the obtained traveled distance cost using our approach drops by 70%. Moreover, for the max tour cost, in FL-MTSP, it changes slowly compared to MDMTSP-GA (Fig. 8.13). The findings demonstrate the effectiveness of our approach in minimizing the overall cost in addition to the max tour cost.

As it is difficult to balance multiple objectives simultaneously, the above results prove that the fuzzy logic system is a good process that allows to combine several conflicting objectives and convert a multiple objective systems into a single objective one.

Also, we prove that the combination of a fuzzy logic system with a heuristic approach leads to optimize the system performance in terms of total traveled distance and max tour cost.

In terms of execution time, it is clear that the gap between our solution and the MDMTSP-GA is very large. The reason for this gap is the fact that the fuzzy logic system from one iteration allows to make decision with no need to repeat the process. This result improves the impact of the use of the fuzzy logic concept that helps to find a solution faster than using only a Heuristic approach (Fig. 8.14).

8.4.4 Comparison with NSGA-II

We also compared FL-MSTP with NSGA-II [15]. NSGA-II is an implementation of multi-objective GA for the MSTP problem. NSGA-II adopted the notion of Pareto optimality to tackle the problem of multi-objective minimization. Suppose we wish to minimize two objectives O_1 and O_2, and let S_i and S_j be two individual solutions.

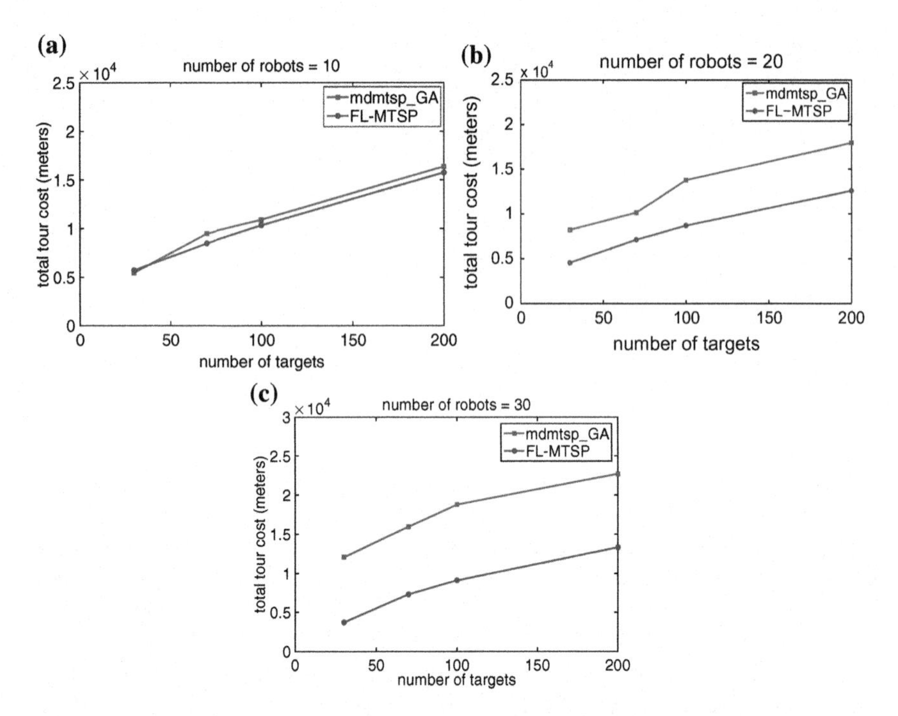

Fig. 8.12 Comparison between FL-MTSP and the MDMTSP_GA in terms of total traveled distance. The results are shown for a different number of targets with a fixed number of robots

Let O_{1i}, O_{2i}, O_{1j}, and O_{2j} be the values of each of the objectives for both solutions. S_i is said to dominate S_j if and only if $((O_{1i} < O_{1j}) \wedge (O_{2i} \leq O_{2j}))$ **OR** $((O_{2i} < O_{2j}) \wedge (O_{1i} \leq O_{1j}))$. The set of non-dominated solutions is called the Pareto-optimal set. Approaches that adopt the notion of Pareto optimality maintain a set of Pareto-optimal solutions from which the decision maker must choose.

NSGA-II description: As any GA, the first step of the NSGA-II is the population initialization step that must be adequate to the problem formulation. Then, the population is sorted based on the non-domination concept. Each individual is given a rank value. Moreover, for each individual, a crowding distance is calculated. The crowding-distance value is calculated as the sum of individual distance values corresponding to each objective [15]. After sorting the population based on the crowded distance and the rank, the best individuals are selected. Next, the crossover and mutation operators are applied to the selected population to generate a child population. The parent population and the child population are combined, sorted based on non-domination, and N individuals are selected based on their crowding distance and their rank. Note that N is the population size. A detailed description of the NSGA-II algorithm is provided in [15].

In our implementation, we used the partial mapped crossover (PMX) operator [16] and the strategy of swapping nodes (belonging to different tours) for mutation. The

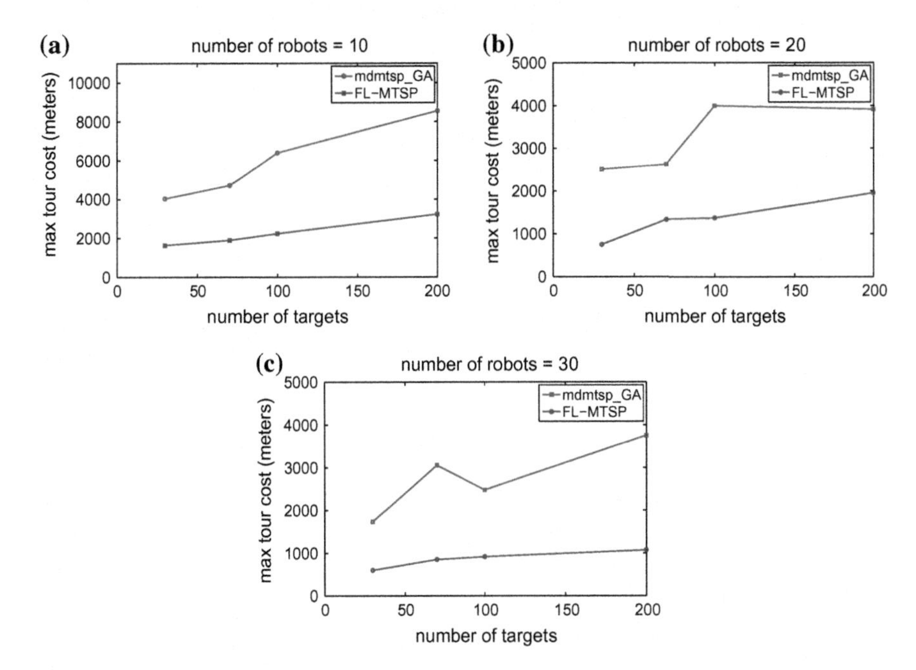

Fig. 8.13 Comparison between FL-MTSP and the MDMTSP_GA in terms of max tour cost. The results are shown for a different number of targets with a fixed number of robots. The number of robots is 10 in **a**, 20 in **b**, and 30 in **c**

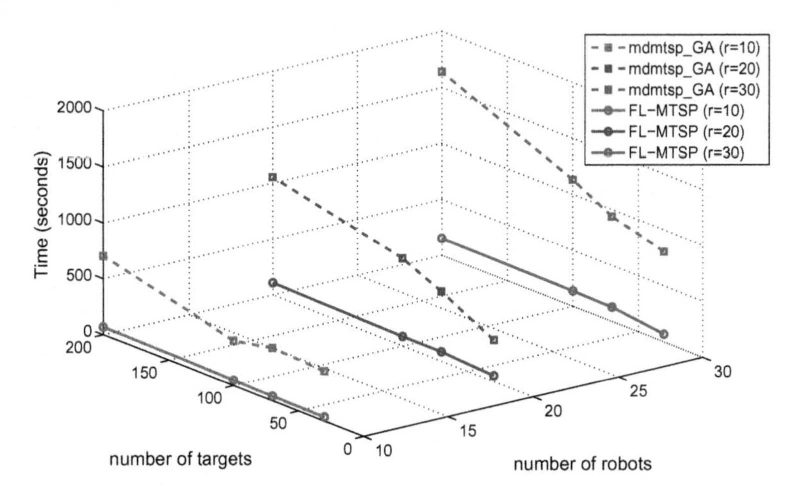

Fig. 8.14 Time comparison between FL-MTSP and the MDMTSP_GA

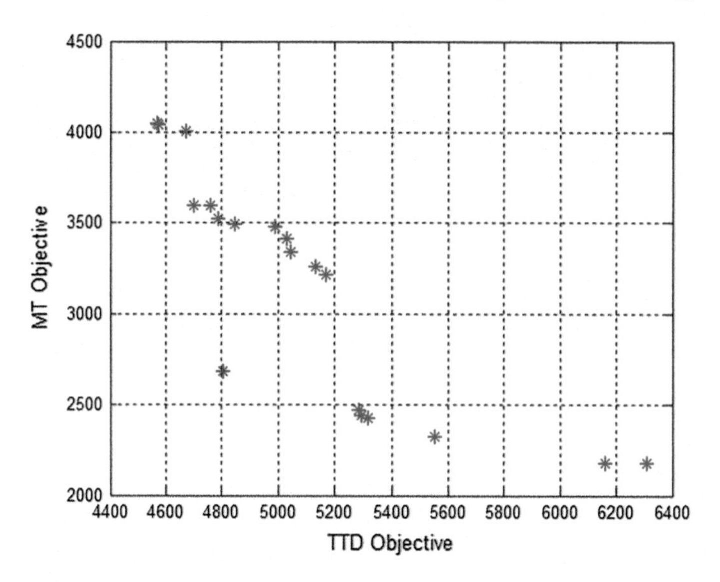

Fig. 8.15 Solutions example obtained for FL-MTSP (blue star) and NSGA-II (red stars)

population size in NSGA-II is set to be 100. The algorithm stops after 50 generations (maximum number of iterations). The crossover probability is equal to 0.7 while the mutation probability is equal to 0.9. We mention that the selection of default parameter values is guided by the simulation results. The simulation scenario for the MD-MTSP consists of 3 robots and 30 targets. Figure 8.15 shows an example of the obtained solutions for both algorithms (FL-MTSP and NSGA-II). The x-axis represents the TTD objective while the y-axis represents the MT objective. The red stars are the solutions generated by the NSGA-II, and the blue star is the solution generated by FL-MTSP.

A set of non-dominated solutions was generated by NSGA-II, and the selection of the best solution depends on the application needs. From the simulation example used for the comparison, we noted that the gap between the solutions of NSGA-II and of FL-MTSP is in the range of [4–24%] for the TTD objective. Also, for the MT objective, the gap varies in [23–34%]. Using the NSGA-II, the solutions are ranged based on their rank of domination. So, most of the solutions will be discarded as they will be assigned lower ranks. This will lead to the loss of promising solutions. Therefore, FL-MTSP provides an acceptable solution in terms of TTD and MT as compared to the NSGA-II. This means that the solution obtained by FL-MTSP is better than at least one of the solutions obtained by NSGA-II at least for one of the objectives.

8.4.5 *Comparison Between FL-MTSP, MDMTSP_GA, MTSP_TT, and MTSP_MT Algorithms*

In this part, we define two new algorithms, namely MTSP_TT, which use the total traveled distance as a metric to assign target locations for the MTSP and the MTSP_MT, which use the max tour cost as a metric to assign target locations for the MTSP. For the MTSP_TT algorithm, we use a greedy algorithm to compute the tour cost for each robot if this target is assigned to it. Then the target will be assigned to the robot that leads to obtain the minimum TTD. The process is repeated until all targets are assigned to the robots. For the MTSP_MT algorithm, we compute the tour cost and select the max tour cost for each robot if this target is assigned to it, using a greedy algorithm. The target will be assigned to the robot with the minimum max tour cost. Also, like the MTSP_TT algorithm, the process is repeated until all targets are assigned to the robots. If the length of the allocated targets list of a robot is higher than m/n, we select the farthest target from its assigned targets and add it to the nearest robot. Then, we apply an existing TSP solver for the tour construction step (Sect. 7.5.2.2). To demonstrate that the proposed fuzzy logic approach provides a good trade-off between MT cost and TTD cost, we performed simulations where we compare our FL-MTSP algorithm MDMTSP_GA [14], MTSP_TT algorithm, and MTSP_MT algorithm. Figure 8.16a and b show the total traveled distance and the max tour cost respectively for the four algorithms. From Fig. 8.16a, we depict that the FL-MTSP algorithm gives better results than the MDMTSP_GA and the MTSP_MT algorithm in terms of total traveled distance. Indeed, when using the max tour cost as a metric, the algorithm optimizes the performance of each robot without considering the benefits of the whole system. This result improves the increase of the total traveled distance for the MTSP_MT algorithm. From Fig. 8.16b, we depict that the FL-MTSP algorithm gives better results than the MDMTSP_GA and the MTSP_TT algorithm in terms of max tour cost. The use of TTD as an optimization criterion leads to increase the max tour cost. We deduced that our FL-MTSP solution proposed to solve the MD-MTSP provides a trade-off between total traveled distance and max tour cost.

8.4.6 *Impact of the TSP Solver on the Execution Time*

To study the impact of the TSP solver,we performed simulations where we used two well known TSP solvers: TSP_GA solver and TSP_LKH solver [11]. The number of robots varies in the interval [3, 10, 20, 30, 100], and the number of target locations was fixed to 200. Figure 8.17 shows the results obtained. It is clearly shown that the execution time of the FL-MTSP algorithm using the TSP-GA solver is more time-consuming than the FL-MTSP algorithm using the TSP-LKH solver. The gap between the FL-MTSP algorithm using the TSP-LKH solver and the FL-MTSP

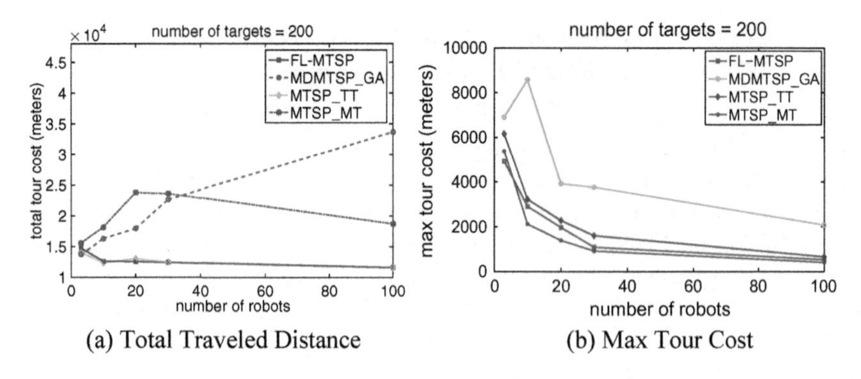

Fig. 8.16 Comparison between FL-MTSP, MDMTSP_GA, MTSP_TT, and MTSP_MT

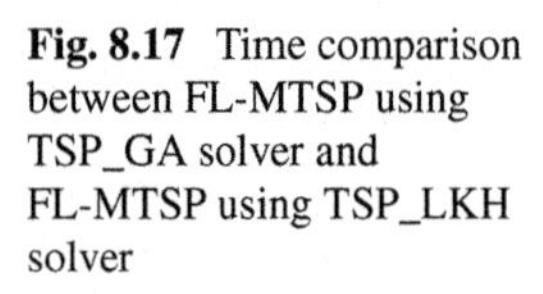
Fig. 8.17 Time comparison between FL-MTSP using TSP_GA solver and FL-MTSP using TSP_LKH solver

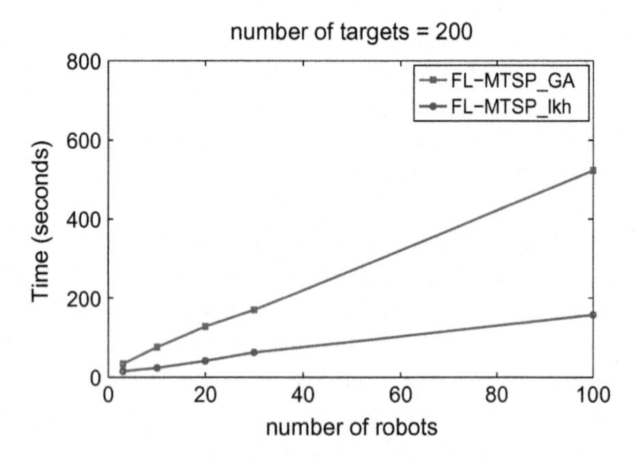

algorithm using the TSP-GA solver increases while increasing the number of robots. Hence, the choice of a good TSP solver helps to improve the execution time of the algorithm while providing a good solution.

8.5 Performance Evaluation of the Move-and-Improve Approach

In this section, we present a realistic simulation study of Move-and-Improve using the Webots simulator [17]. Webots is a professional mobile robot simulator that provides the possibility to create a 3D virtual worlds that model with high accuracy the physical properties, dynamics and kinematics of the mobile robots, and the objects surrounding the environment. In Webots, a simulation model consists in defining the virtual physical worlds, the robots, and the controllers of the robots that implement their behavior. In our simulation model, we used a rectangular *checkered-marble* arena provided by default in Webots and a realistic model of the Pioneer 3AT robot. It has four wheels and is equipped with 16 sonar distance sensors [18]. We have developed two types of controllers in Webots in the C++ language. The first is a robot controller that determines the operations of a robot and implements the Move-and-Improve algorithm. The second is a supervisor controller, which is responsible for generating targets, sending them to the robots, collecting and displaying statistics, and the simulation evolution. A video demonstration that illustrates an example of a simulation scenario is available at [19]. The code can be downloaded from the iroboapp Web site [10].

System Settings. We considered a simulation world comprising 30 targets deployed in an area of 40 m $\times$ 40 m with some obstacles. We evaluated the Move-and-Improve mechanism under several scenarios with different configurations of the number of robots 3, 5, and 7 robots and communication ranges (from 2 to 20 m). Each scenario is characterized by a specified fixed number of robots, a fixed number of targets and a robot communication range. Each simulation configuration is repeated 10 times and for each simulation run. The values presented in the following figures represent the average of the collected values. The communication between the robots was achieved using wireless radio links.

We considered three evaluation metrics: (1) the total traveled distance, (2) the Maximum Tour (i.e., MTD), and (3) the communication overhead, which represents of the number of messages exchanged between the cooperative robots.

Figure 8.18 presents the total traveled distance for different values of the communication range. It is clear from the figure that the TTD decreases when the communication range increases. This is expected as a larger communication range allows the robots to avoid overlapped targets and thus reduces the distance they will have to move. The slope of TTD decreases sharper for shorter communication ranges, namely from 2 to 8 m. Increasing the communication range above 8 m resulted in a smoother decrease of the TTD, in particular with high robot density (i.e., case of 7 robots).

Figure 8.19 presents the communication overhead (as number of messages sent and received) for different values of communication range. We noted from figure that large value of communication range leads to high communication overhead. Moreover, for large value of communication range the communication overhead exponentially increases with number of robots.

Figure 8.20 presents the ratio of overlapped targets as function of the communication range. It is clear from the figure that the ratio of overlapped targets tends to 0 when the communication range tends to 20, which represents approximately the half of the maximum distance between two robots in the simulated world (40 m $\times$ 40 m). This result is reasonable as with enough communication between robots common targets will be eliminated.

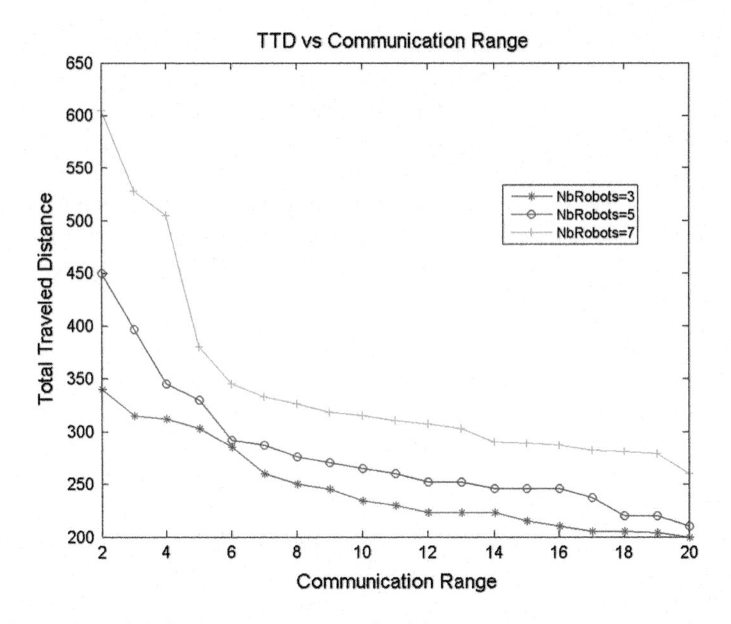

Fig. 8.18 Total traveled distance versus communication range

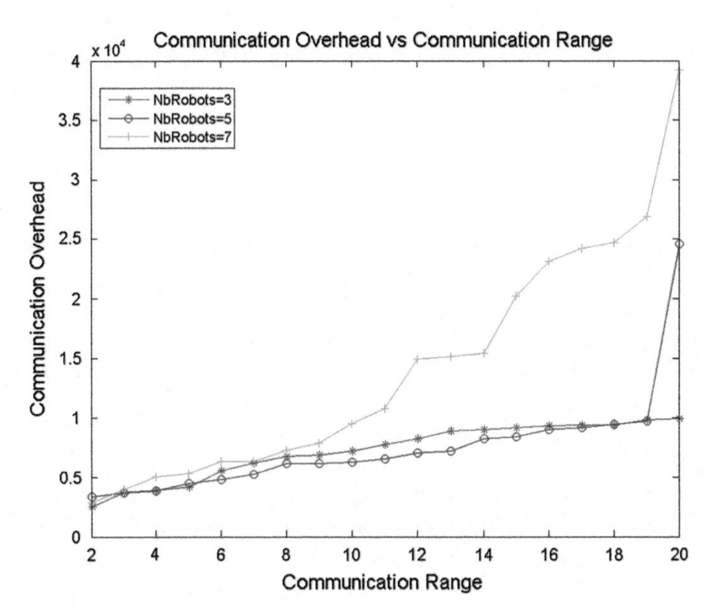

Fig. 8.19 Communication overhead versus communication range

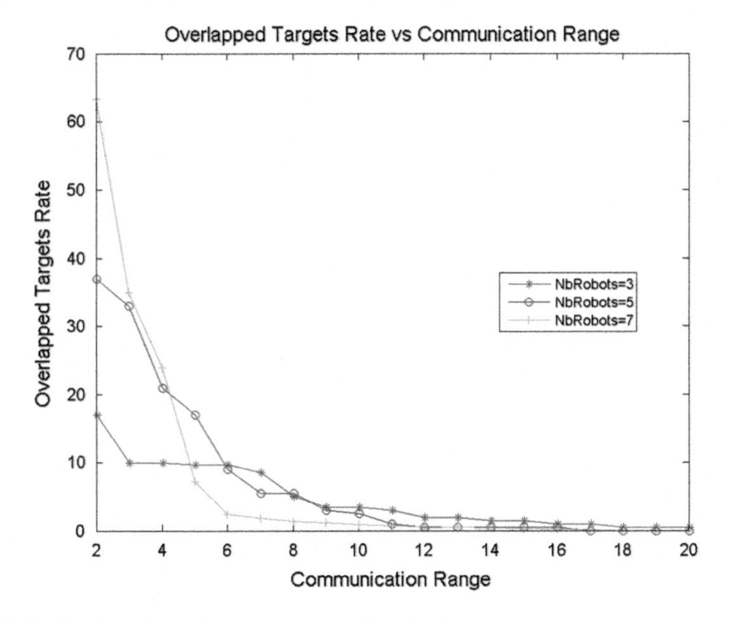

Fig. 8.20 Ratio of overlapped targets versus communication range

8.6 Conclusion

In this chapter, we have presented the performance evaluation of the four proposed approaches for solving the MRTA problem.

The IDMB approach consists in allocating each task to a robot then swapping tasks to ameliorate the cost of the assignment. It has been proved that the IDMB algorithm produces near-optimal solutions, and in several cases, it gives the optimal solution.

The CM-MTSP is based on the use of a clustering method with an auction process. The algorithm has been compared against a single objective and greedy algorithms. It has been concluded that the the CM-MTSP approach provides a good trade-off between the objectives, as compared to a single-objective algorithm with an improvement of the mission time and reduces the execution time as compared to the greedy algorithm.

The FL-MTSP approach is a centralized solution based on the use of the fuzzy logic algebra to combine objectives. This approach has been compared against an existing MD-MTSP solver based on the genetic algorithm. The approach outperformed the GA approach on both the objectives, the total traveled distance and the maximum traveled distance, and also in terms of execution time. We compared the solution against two single-objective algorithms. We proved that our multi-objective algorithm provides a trade-off between the total traveled distance and the maximum traveled distance.

The Move-and-Improve approach is a distributed approach where robots communicate together in order to produce optimal allocation of targets. From the simulation, we noticed that as the communication range of robots increases, the total traveled distance and the ratio of overlapped targets decrease but the communication overhead increases. This result proved that Move-and-Improve algorithm can produce optimal or near-optimal solutions.

References

1. Viguria, Antidio, and Ayanna M. Howard. 2009. An integrated approach for achieving multi-robot task formations. *IEEE/ASME Transactions on Mechatronics* 14 (2): 176–186.
2. Ann Shim, Vui, KC Tan, and CY Cheong. 2012. A hybrid estimation of distribution algorithm with decomposition for solving the multiobjective multiple traveling salesman problem. *IEEE Transactions on Systems, Man, and Cybernetics, Part C (Applications and Reviews)*, 42(5): 682–691.
3. Ke, Liangjun, Qingfu Zhang, and Roberto Battiti. 2013. Moea/d-aco: A multiobjective evolutionary algorithm using decomposition and antcolony. *IEEE Transactions on Cybernetics* 43 (6): 1845–1859.
4. Cheikhrouhou, Omar, Anis Koubâa, and Hachemi Bennaceur. 2014. Move and improve: A distributed multi-robot coordination approach for multiple depots multiple travelling salesmen problem. In *2014 IEEE international conference on autonomous robot systems and competitions (ICARSC)*, 28–35. IEEE.
5. Kivelevitch, Elad, Kelly Cohen, and Manish Kumar. 2013. A market-based solution to the multiple traveling salesmen problem. *Journal of Intelligent and Robotic Systems*: 1–20.
6. Miettinen, Kaisa. 2012. *Nonlinear multiobjective optimization*, vol. 12. Berlin: Springer Science & Business Media.
7. W Kuhn, Harold. 1955. The hungarian method for the assignment problem. *Naval Research Logistics (NRL)* 2(1-2): 83–97.
8. Ackerman, Evan. 2012. TurtleBot. http://www.turtlebot.com/. Accessed on Oct 2012.
9. Robot Operating System (ROS). http://www.ros.org.
10. Koubaa, Anis. 2014. The Iroboapp Pproject. http://www.iroboapp.org. Accessed 27 Jan 2016.
11. Helsgaun, Keld. 2012. Lkh. http://www.akira.ruc.dk/~keld/research/LKH/.
12. Alexis, Kostas, Georgios Darivianakis, Michael Burri, and Roland Siegwart. 2016. Aerial robotic contact-based inspection: Planning and control. *Autonomous Robots* 40 (4): 631–655.
13. Yong, Wang. 2015. Hybrid max-min ant system with four vertices and three lines inequality for traveling salesman problem. *Soft Computing* 19 (3): 585–596.
14. Kivelevitch, Elad. 2011. Mdmtspv_ga - multiple depot multiple traveling salesmen problem solved by genetic algorithm. http://www.mathworks.com/matlabcentral/fileexchange/31814-mdmtspv-ga-multiple-depot-multiple-traveling-salesmen-problem-solved-by-genetic-algorithm.
15. Deb, Kalyanmoy, Amrit Pratap, Sameer Agarwal, and T.A.M.T. Meyarivan. 2002. A fast and elitist multiobjective genetic algorithm: Nsga-ii. *IEEE Transactions on Evolutionary Computation* 6 (2): 182–197.
16. Bolaños, R., M. Echeverry, and J. Escobar. 2015. A multiobjective non-dominated sorting genetic algorithm (nsga-ii) for the multiple traveling salesman problem. *Decision Science Letters* 4 (4): 559–568.
17. Webots: the mobile robotics simulation software. 2014. http://www.cyberbotics.com/.
18. Pioneer3at robots. 2014. http://www.mobilerobots.com/ResearchRobots/P3AT.aspx.
19. Webots simulation scenarios. 2014. http://www.iroboapp.org/index.php?title=Videos.

Index

A. Koubaa et al., *Robot Path Planning and Cooperation*, Studies in Computational Intelligence 772, https://doi.org/10.1007/978-3-319-77042-0

Printed by Printforce, the Netherlands